BC SCIENCE

Student Workbook

Authors

Briar Ballou
Handsworth Secondary School
North Vancouver, British Columbia

Van Chau
Delview Secondary School
Delta, British Columbia

Christine Weber
Science Writer
Chilliwack, British Columbia

Program Consultants

Sandy Wohl
Hugh Boyd Secondary School
Richmond, British Columbia

Herb Johnston
Faculty of Education
University of British Columbia
Vancouver, British Columbia

 **McGraw-Hill Ryerson**

Toronto   Montréal   Boston   Burr Ridge, IL   Dubuque, IA
Madison, WI   New York   San Francisco   St. Louis   Bangkok   Bogotá
Caracas   Kuala Lumpur   Lisbon   London   Madrid   Mexico City
Milan   New Delhi   Santiago   Seoul   Singapore   Sydney   Taipei

MAY BE OBTAINED BY
CONTACTING:

McGraw-Hill Ryerson Ltd.

WEB SITE:
http://www.mcgrawhill.ca

E-MAIL:
orders@mcgrawhill.ca

TOLL-FREE FAX:
1-800-463-5885

TOLL-FREE CALL:
1-800-565-5758

OR BY MAILING YOUR
ORDER TO:
McGraw-Hill Ryerson
Order Department
300 Water Street
Whitby, ON L1N 9B6

Please quote the ISBN and
title when placing your order.

McGraw-Hill Ryerson
BC Science 10 Workbook

ISBN-13: 978-0-07-098461-5
ISBN-10: 0-07-098461-1

http://www.mcgrawhill.ca

9 MP 1 9 8 7 6 5 4 3 2

Printed and bound in Canada

PUBLISHER: Diane Wyman
DEVELOPMENT HOUSE EDITORS: First Folio Resource Group, Inc.
DEVELOPMENTAL EDITORS: Kendall Anderson, Julie Kretchman
MANAGER, EDITORIAL SERVICES: Crystal Shortt
SUPERVISING EDITOR: Janie Deneau
COPY EDITOR: Kendall Anderson
EDITORIAL ASSISTANT: Michelle Malda
MANAGER, PRODUCTION SERVICES: Yolanda Pigden
PRODUCTION CO-ORDINATOR: Sheryl MacAdam
COVER DESIGN: Valid Layout & Design
ART DIRECTION: Laserwords
ELECTRONIC PAGE MAKE-UP: Laserwords

Contents

UNIT 3 Motion

UNIT 4 Energy Transfer in Natural Systems

© 2008 McGraw-Hill Ryerson Limited

Chapter 11 Climate change occurs through natural processes and human activities.

Chapter 12 Thermal energy transfer drives plate tectonics.

Biomes

Textbook pages 8–33

Before You Read

A biome includes large regions that have similar living and non-living components. Tundra and desert are two examples of biomes. What other biomes can you name?

_____

_____

 *Make Flash Cards*

For each biome, write a question on one side of the flash card. Write the answer on the other side. Quiz yourself until you can answer all the questions.

What is a biome?

The biosphere is the thin layer of air, land, and water on or near Earth's surface where living things exist. A **biome** is the largest division of the biosphere. Biomes are characterized by their **biotic** (living) and **abiotic** (non-living) components.

What are Earth's biomes like?

Earth has eight terrestrial (land-based) biomes.

Biome	Main characteristics
tundra	■ located in the upper northern hemisphere; very cold and dry ■ due to permanently frozen soil, plants are short and there are few trees
boreal forest	■ found in the far north; below freezing half the year ■ mainly coniferous (cone-bearing) trees
temperate deciduous forest	■ located in temperate regions, mostly eastern North America, eastern Asia, and western Europe ■ trees lose their leaves in winter ■ large seasonal changes with four distinct seasons
temperate rainforest	■ found along coastlines where ocean winds drop large amounts of moisture ■ cool and very wet, allowing trees (mainly evergreens) to grow very tall
grassland	■ occurs in temperate and tropical regions ■ covered with grasses that have deep roots, which are well adapted for drought
tropical rainforest	■ found in a wide band around the equator ■ wet and warm year-round, allowing for the growth of a dense canopy of tall trees
desert	■ occur in temperate and tropical regions; days are hot and nights are cold ■ rainfall is minimal and plants and animals are adapted to reduce water loss
permanent ice	■ includes the polar land masses and large polar ice caps ■ the few animals that live here are well insulated against the extreme cold

How do abiotic factors influence the characteristics and distribution of biomes?

Identical biomes are found in different parts of the world. These biomes all have similar plants and animals (biotic factors) because they have similar temperatures and precipitation patterns (abiotic factors). Temperature and precipitation are the main abiotic factors that influence the characteristics and distribution of biomes. The following factors influence temperature and precipitation:
- ◆ latitude
- ◆ elevation
- ◆ wind
- ◆ ocean currents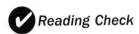

What are climatographs?

Climate is the average pattern of weather conditions that occur in a region over a period of years. Temperature and precipitation are two important factors that determine climate. A **climatograph** is a graph of climate data for a specific region. A climatograph gives average temperature and average total precipitation for each month.

How are organisms adapted to the specific conditions of their biome?

Adaptations are characteristics that enable organisms to better survive and reproduce. Organisms are specially adapted for survival in the specific environmental conditions of their biome. There are three types of adaptations:
- ◆ **structural adaptations**: physical parts or features of an organism that enable it to survive and reproduce.
- ◆ **physiological adaptations**: a chemical or physical event that takes place in the body of an organism to support its ability to survive and reproduce
- ◆ **behavioural adaptations**: things that an organism does (ways that it behaves) that enable it to survive and reproduce.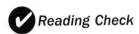

✔ *Reading Check*

What are the two main abiotic factors that influence the character-istics and distribution of biomes?

_____

_____

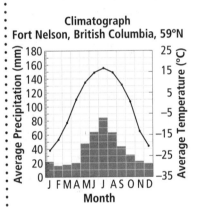

Climatograph
Fort Nelson, British Columbia, 59°N

✔ *Reading Check*

List the three main types of adaptations displayed by organisms.

_____

_____

_____

Use with textbook pages 8–28.

Biomes and ecosystems

Vocabulary	
abiotic	latitude
adaptations	ocean currents
behavioural	physiological
biome	precipitation
biotic	structural
climatograph	temperature
elevation	terrestrial

Use the terms in the vocabulary box to fill in the blanks. Use each term only once.

1. _____ components are the living organisms in an environment, such as plants, animals, fungi, and bacteria.

2. _____ components are the non-living parts of an environment, such as sunlight, soil, moisture, and temperature.

3. A _____ includes large regions that have similar biotic components and abiotic components.

4. A _____ biome is land-based.

5. _____ and _____ are two important abiotic factors that influence the characteristics of biomes and the distribution of biomes on Earth.

6. _____ is the distance measured in degrees north or south from the equator.

7. _____ is the height of a land mass above sea level.

8. _____ are another abiotic factor that affects temperature and precipitation and therefore influences the characteristics of biomes.

9. A _____ is a graph of climate data for a specific region and is generated from data usually obtained over 30 years from local weather observation stations.

10. _____ are characteristics that enable organisms to better survive and reproduce.

11. A _____ adaptation is a physical feature of an organism's body having a specific function that contributes to the survival of the organism. A _____ adaptation is a physical or chemical event that occurs within the body of an organism that enables survival. A _____ adaptation refers to what an organism does to survive in the unique conditions of its environment.

Use with textbook pages 20–28.

Various biomes

Complete the following table, describing the general locations and two to three main physical features of the eight terrestrial biomes.

Biome	Location(s)	Physical features
tundra		
boreal forest		
temperate deciduous forest		
temperate rainforest		
grassland (temperate and tropical)		
tropical rainforest		
desert (hot and cold)		hot desert: cold desert:
permanent ice (polar ice)		

Use with textbook pages 16–28.

Climatographs

Which world biomes are represented by the data in the following climatographs?

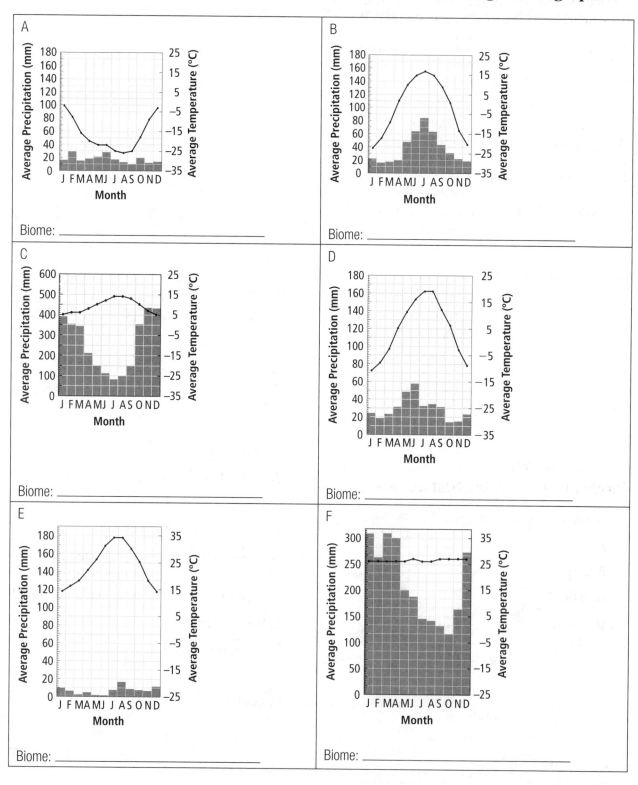

A

Biome: _____

B

Biome: _____

C

Biome: _____

D

Biome: _____

E

Biome: _____

F

Biome: _____

Use with textbook pages 8–28.

Biomes

Match each Term on the left with the best Descriptor on the right. Each Descriptor may only be used once.	
Term	**Descriptor**
1. _____ abiotic 2. _____ adaptations 3. _____ biome 4. _____ biotic 5. _____ climate 6. _____ latitude	**A.** the distance measured in degrees north or south from the equator **B.** characteristics that enable organisms to better survive and reproduce **C.** relating to non-living parts of an environment, such as sunlight, soil, moisture, and temperature **D.** relating to the living organisms, such as plants, animals, fungi, and bacteria **E.** the largest division of the biosphere **F.** the average conditions of the atmosphere in a large region over 30 years

Circle the letter of the best answer.

7. A biome is best represented by a:

A. river

B. city

C. latitude

D. desert

8. Which of the following is an abiotic component of an environment?

A. algae

B. sunlight

C. fungi

D. plants

9. Which of the following is a biotic component of an ecosystem?

A. moisture

B. sand

C. bacteria

D. temperature

10. Which of the following is a characteristic of the boreal forest biome?

A. below freezing half the year

B. long, hot summers

C. polar land masses

D. lots of precipitation

11. Which world biome is represented by a climatograph that illustrates an average precipitation of 300 cm in the month of January?

A. grassland

B. tropical rainforest

C. permanent ice

D. temperate deciduous forest

12. Which world biome is represented by a climatograph that illustrates an average temperature of –25°C in the month of July?

A. boreal forest

B. tropical rainforest

C. permanent ice

D. tundra

Ecosystems

Textbook pages 34–51

Before You Read

How do you think ecosystems are related to the biomes you learned about in the previous section? Record your ideas below.

? Create a Quiz

Create a quiz to help you learn the boldface terms introduced in this section. Answer your questions and share your quiz with your classmates.

✔ Reading Check

What are the two main components of an ecosystem?

_____

_____

✔ Reading Check

Organize the following in the correct ecological hierarchy: community, ecosystem, species, population.

_____

_____

_____

_____

What is an ecosystem?

In an **ecosystem**, abiotic components, such as oxygen, water, nutrients, light, and soil, support the life functions of biotic components, such as plants, animals, and micro-organisms. Biomes contain many different ecosystems. Ecosystems can be small. Examples of small ecosystems include a tide pool and a rotting log. Ecosystems also can be large. Examles of large ecosystems include a coastal Douglas fir forest and a biome.

Ecosystems contain different habitats. A **habitat** is the place in which an organism lives. For example, a sculpin is a fish that makes its habitat between rocks at the bottom of a tide pool ecosystem. ✔

How are biotic interactions in ecosystems structured?

Organisms within an ecosystem constantly interact to obtain resources, such as food, water, sunlight, or habitat. As a result of these interactions, organisms have special roles—or **niches**—in their ecosystems. An organism's niche includes the way in which it contributes to and fits into its environment. Many different organisms can live in the same habitat if they occupy different niches. Biotic interactions are structured from smallest to largest in an **ecological hierarchy**.

♦ A **species** is a group of closely related organisms that can reproduce with one another.

♦ All the members of a species within an ecosystem are referred to as a **population**.

♦ Populations of different species that interact in a specific ecosystem form a **community**. ✔

What different biotic interactions occur in ecosystems?

Symbiosis refers to the interaction of two different organisms that live in close association. **Commensalism, mutualism, and parasitism** are types of symbiotic interactions. Other biotic interactions include **competition, predation,** and **mimicry**.

Interaction	Result	Example
commensalism	One organism benefits and the other organism is neither helped nor harmed.	Barnacles attach to whales and are transported to new locations in the ocean.
mutualism	Both organisms benefit and sometimes neither species can survive without the other.	In lichen, the alga produces sugars and oxygen for the fungus, which provides carbon dioxide and water for the alga.
parasitism	One species benefits and another is harmed.	Hookworms attach to the gut wall and obtain nourishment from their host's blood.
competition	Organisms require the same resource (such as food) in the same location at the same time.	Spotted knapweed releases chemicals into the soil, which prevents the growth of other plants.
predation	One organism (the predator) eats all or part of another organism (the prey).	Cougars have sharp, pointed teeth to catch prey.
mimicry	Prey animal mimics another species that is dangerous or tastes bad to avoid being eaten.	Viceroy butterflies look like bitter-tasting monarch butterflies and are avoided by predators.

Name ______________________ Date ______________________

Comprehension
Section 1.2

Use with textbook pages 34–48.

Parts of an ecosystem

1. What is the difference between an ecosystem and a habitat?

____

____

2. List three main abiotic components of ecosystems.

____

____

3. What is the difference between a population and a community?

____

____

4. Define the term symbiosis.

____

____

5. What is commensalism?

____

____

6. How does mutualism differ from parasitism?

____

____

7. What is predation?

____

____

Use with textbook pages 39–47.

Biotic interactions in ecosystems

Vocabulary
biosphere   organism
community   population
ecosystem

1. Use the vocabulary words in the box above to label the Williams Creek ecosystem.

2. Give the ecological hierarchy for these biotic interactions from largest to smallest.

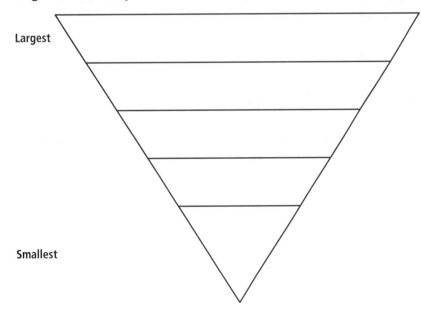

Largest

Smallest

3. List three populations that interact in your community.

_____

_____

Use with textbook pages 39–47.

Symbiotic relationships

- commensalism
- mutualism

- parasitism
- competition

- predation

Use the above terms to identify the following descriptions. Identify the term and explain the relationship.

1. An ant lives in the thorns of the bullhorn acacia bush. The ant sips the nectar of the acacia's leaflets. The ants protect the plant by fighting off other insects.

   Term: _____

   Explanation: _____

   _____

2. Spotted knapweed releases chemicals into the soil. These chemicals prevent the growth of other plants and allow the plant to spread quickly.

   Term: _____

   Explanation: _____

   _____

3. Lynx hunt snowshoe hares. When the lynx population increases the snowshoe hare population decreases.

   Term: _____

   Explanation: _____

   _____

4. Spanish moss lives on trees in rainforests and has no roots. The feathery structure of the Spanish moss captures nutrients and moisture from the air.

   Term: _____

   Explanation: _____

   _____

5. The mountain pine beetle is killing British Columbia's lodgepole and white pine forests.

   Term: _____

   Explanation: _____

   _____

Use with textbook pages 34–48.

Ecosystems

Match each Term on the left with the best Descriptor on the right. Each Descriptor may be used only once.	
Term	**Descriptor**
1. _____ commensalism 2. _____ competition 3. _____ ecosystem 4. _____ mutualism 5. _____ niche 6. _____ parasitism 7. _____ predation	**A.** the special role an organism plays in an ecosystem **B.** a part of a biome in which abiotic components interact with biotic components **C.** a symbiotic relationship in which one species benefits and another is harmed **D.** a symbiotic relationship in which one species benefits and the other species is neither helped nor harmed **E.** a harmful interaction between two or more organisms that occurs when the organisms compete for the same resource in the same location at the same time **F.** a symbiotic relationship between two organisms in which both organisms benefit **G.** predator-prey interactions in which one organism eats all or part of another organism

Circle the letter of the best answer.

8. What relationship is demonstrated by a barnacle being attached to a whale?

  A. mutualism

  B. commensalism

  C. parasitism

  D. competition

9. Which of the following is an example of mutualism?

  A. similar colouring of shrimp and crimson anemone

  B. hookworms attaching to a dog's intestine

  C. coyotes hunting in packs to kill large animals

  D. snapdragon flowers that open for bees of a specific mass

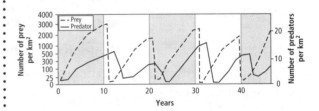

10. Which of the following situations best describes the relationship of the predator and prey population shown in the graph above?

  A. As the predator population increases the prey population increases.

  B. As the predator population decreases the prey population decreases.

  C. As the predator population increases the prey population decreases.

  D. Predator population has no effect on prey population.

Energy Flow in Ecosystems

Textbook pages 56–67

Before You Read

In this section, you will explore food chains and food webs, as well as food pyramids. What are the main differences between a chain and a web? Record your ideas below.

_____

_____

 Mark the Text

In Your Own Words

Define the bold terms in this summary in your own words.

✓ Reading Check

1. What are the different steps in a food chain called?

_____

_____

How does energy flow in an ecosystem?

Energy flow is the transfer of energy from one organism to another in an ecosystem. Every organism interacts with its ecosystem in two ways:

1. the organism obtains food energy from the ecosystem

2. the organism contributes energy to the ecosystem

How are energy flow and feeding relationships in ecosystems modelled?

Ecologists use three models to illustrate energy flow and feeding relationships in an ecosystem:

1. Food chains: **Food chains** show the flow of energy from plant to animal and from animal to animal. Plants are called **producers** because they "produce" food in the form of carbohydrates during photosynthesis. **Consumers** eat plants and other organisms. Each step in a food chain is called a **trophic level.** ✓

Trophic Level	Type of Organism	Energy Source	Example
1st	**primary producer**	obtain energy from the Sun	grass, algae (plants)
2nd	**primary consumer**	obtain energy from primary producers	grasshoppers, krill **(herbivores)**
3rd	**secondary consumer**	obtain energy from primary consumers	frogs, crabs **(carnivores)**
4th	**tertiary consumer**	obtain energy from secondary consumers	hawks, sea otters (top carnivores)

2. Food webs: Many animals are part of more than one food chain in an ecosystem because they eat or are eaten by several organisms. Interconnected food chains are illustrated in a model called a **food web.** Animals that eat plants and other animals are called omnivores.

3. Food pyramids: A **food pyramid** (or **ecological pyramid**) is a model that shows the loss of energy from one trophic level to another. When one organism consumes another, the energy stored in the food organism is transferred to the consumer. However, not all of this energy is incorporated into the consumer's tissues. Between 80 and 90 percent of it is used for chemical reactions and is lost as heat. This means ecosystems can support fewer organisms at higher trophic levels, as less energy reaches these levels. ✔

great horned
owl

long-tailed
weasel

jackrabbit

grasses

Food pyramid

✔ **Reading Check**
1. What does a food pyramid demonstrate?

_____

_____

_____

_____

_____

How do dead organisms contribute to energy flow?

Decomposition describes the breakdown of organic wastes and dead organisms. Energy is released in decomposition. When living organisms carry out decomposition, it is called **biodegradation**.

◆ **Detrivores,** such as small insects, earthworms, bacteria, and fungi, obtain energy and nutrients by eating dead plants and animals, as well as animal waste.

◆ **Decomposers,** such as bacteria and fungi, change wastes and dead organisms into nutrients that can once again be used by plants and animals.

Detrivores and decomposers feed at every trophic level.

Use with textbook pages 56–64.

Energy flow

Vocabulary	
biodegradation	food webs
biomass	photosynthesis
consumer	primary consumers
decomposers	primary producers
decomposition	secondary consumers
energy flow	tertiary consumers
food chains	trophic
food pyramids	

Use terms in the vocabulary box to fill in the blanks. Use each term only once.

1. _____ refers to the total mass of living plants, animals, fungi, and bacteria in a given area.

2. The flow of energy from an ecosystem to an organism and from one organism to another is called _____.

3. Plants are called producers because they "produce" food in the form of carbohydrates during _____

4. An insect, such as a bee, that feeds on a plant is called a _____.

5. _____ is the breaking down of organic wastes and dead organisms.

6. The action of living organisms, such as bacteria, to break down organic matter is called _____.

7. _____ change waste and dead organisms into usable nutrients.

8. _____ are models that show the flow of energy from plant to animal and from animal to animal. Each step is called a _____ level.

9. Plants and phytoplankton, such as algae, are at the first trophic level and are referred to as _____.

10. _____ obtain their energy from primary producers.
_____ obtain their energy by eating primary consumers.

11. In the fourth trophic level are _____ that feed on secondary consumers to obtain energy.

12. _____ are models of the feeding relationships within an ecosystem.
_____ show the loss of energy from one trophic level to another.

Use with textbook pages 60–64.

Food chains, food webs, and food pyramids

Use the diagrams to help you answer the questions.

Scientific model	Questions
food chain red-tailed hawk spotted frog grasshoper bunchgrass Terrestrial food chain sea otter crab krill algae Aquatic food chain	1. What plants or animals are the primary producers in this food chain? _____ _____ 2. What trophic level do the frogs and crabs belong to? _____ 3. What do tertiary consumers feed on to obtain energy? _____
food web grizzly bear red-tailed hawk chipmunk deer grouse marmot decomposers and detrivores berries and flowers grasses seeds	4. What term is used to describe a chipmunk that eats seeds or fruit? _____ 5. What kind of consumers do omnivores eat? _____ 6. Give two examples of detrivores. _____ _____

food pyramid

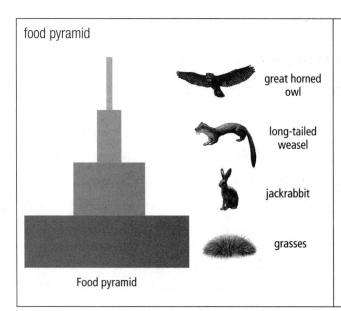

great horned
owl

long-tailed
weasel

jackrabbit

grasses

Food pyramid

7. What is a food pyramid?

_____

_____

8. Which level of a food pyramid stores the most
energy?

_____

9. Which level of a food pyramid stores the least amount
of energy?

_____

Name Date

Illustrating Concepts

Section 2.1

Use with textbook pages 58–64

Modelling a local ecosystem

Reflect on the plants and animals that exist in your local ecosystem.

1. List 12 plants and animals. Remember to represent each of the trophic levels.

2. Organize four of these plants and animals into a food chain.

3. Using all 12 of the plants and animals, design a food web that illustrates the feeding relationships within your selected ecosystem.

4. Organize the plants and animals into a food pyramid that demonstrates the loss of energy as you move from one trophic level to the next.

Use with textbook pages 56–64.

Energy flow in ecosystems

Match each Term on the left with the best Descriptor on the right. Each Descriptor may only be used once.	
Term	**Descriptor**
1. _____ biodegradation 2. _____ consumers 3. _____ decomposers 4. _____ food chain 5. _____ food pyramid 6. _____ food web 7. _____ producers 8. _____ trophic levels	**A.** a model that shows the flow of energy from plant to animal and from animal to animal **B.** organisms that produce food in the form of carbohydrates during photosynthesis **C.** the breaking down of dead organic matter by organisms, such as bacteria **D.** steps in a food chain that show feeding and niche relationships among organisms **E.** a model that shows the loss of energy from one trophic level to another **F.** an organism that eats other organisms **G.** a model of the feeding relationships within an ecosystem **H.** organisms that break down wastes and dead organisms and change them into usable nutrients

Circle the letter of the best answer.

9. In a food chain, primary producers are usually:

   A. amphibians **C.** mammals

   B. bacteria **D.** plants

10. What product of photosynthesis supplies energy for life forms?

    A. carbohydrates

    B. carbon dioxide

    C. oxygen

    D. water

11. Which of the following organisms are likely to be found in the third trophic level of a food chain?

    A. algae **C.** grasshopper

    B. frog **D.** hawk

12. Which of the following describes the process of biodegradation?

    A. plants using photosynthesis to create food

    B. primary consumers eating plants

    C. bacteria breaking down organic matter

    D. omnivores eating plants and animals

13. In a food pyramid, how much energy is lost from trophic level to trophic level?

    A. 20 % **C.** 70%

    B. 50 % **D.** 90%

14. In a food pyramid:

    A. as the trophic level decreases, the number of organisms supported by the ecosystem decreases

    B. as the trophic level increases, the number of organisms supported by the ecosystem increases

    C. as the trophic level increases, the number of organisms supported by the ecosystem stays the same

    D. as the trophic level decreases, the number of organisms supported by the ecosystem increases

Nutrient Cycles in Ecosystems

Textbook pages 68–91

Before You Read

Like other organisms, your body relies on nutrients to stay healthy. Based on your current understanding, create a definition of what you think a nutrient is. Write your definition in the lines below.

_____

_____

 Mark the Text

Check for Understanding
As you read this section, be sure to reread any parts you do not understand. Highlight any sentences that help you improve your understanding.

✔ **Reading Check**

1. Name the three main nutrient cycles.

_____

_____

_____

How are nutrients cycled in the biosphere?

Nutrients are chemicals required for plant and animal growth and other life processes. They are constantly recycled within Earth's biosphere. Nutrients spend different amounts of time in **stores** within the atmosphere, oceans, and land. Nutrients are stored for short periods of time in short-term stores, such as living organisms and the atmosphere. Nutrients can also be incorporated into longer-term stores, such as Earth's crust. **Nutrient cycles** describe the flow of nutrients in and out of stores as a result of biotic and abiotic processes. Without human interference, nutrient cycles are almost perfectly balanced. There are three main cycles that move nutrients through terrestrial and aquatic ecosystems:

1. the **carbon cycle**

2. the **nitrogen cycle**

3. the **phosphorus cycle** ✔

How does the carbon cycle work?

Carbon is an essential component of cells and life-sustaining chemical reactions. Carbon is cycled through living and decaying organisms, the atmosphere, bodies of water, and soil and rock. Carbon moves between stores via six main processes:

1. Photosynthesis: **Photosynthesis** is a chemical reaction that converts solar energy and atmospheric carbon dioxide gas (CO_2) into chemical energy.

2. Cellular respiration: During **cellular respiration**, plants and animals obtain energy by converting carbohydrates and oxygen (O_2) into carbon dioxide and water.

Name

Date

Section

2.2

Summary

continued

3. Decomposition: Decomposers release carbon dioxide into the atmosphere through the **decomposition** of carbon-rich organic matter in soil.

4. Ocean processes: Dissolved carbon dioxide is stored in oceans. Marine organisms store carbon-rich **carbonate** (CO_3^{2-}) in their shells, which eventually form sedimentary rock.

5. Volcanic eruptions

6. Forest fires

How do human activities affect the carbon cycle?

Human activities, such as fossil fuel combustion and land clearance, quickly introduce carbon into the atmosphere from longer-term stores. These actions increase the levels of carbon dioxide, a greenhouse gas that contributes to global climate change.

How does the nitrogen cycle work?

Nitrogen is an important component of DNA and proteins. Most nitrogen is stored in the atmosphere, where it exists as nitrogen gas (N_2). It is also stored in bodies of water, living organisms, and decaying organic matter. Most organisms cannot use atmospheric nitrogen gas. The nitrogen cycle involves four processes, three of which make nitrogen available to plants and animals.

1. **Nitrogen fixation:** Nitrogen gas is converted into nitrate (NO_3^-) and ammonium (NH_4^+), compounds that are usable by plants. Nitrogen fixation occurs mainly through nitrogen-fixing bacteria, and when lightning strikes in the atmosphere.

2. **Nitrification:** Ammonium is converted into nitrate and nitrite (NO_2^-) through the work of **nitrifying bacteria**.

3. **Uptake**: Useable forms of nitrogen are taken up by plant roots and incorporated into plant proteins. When herbivores and omnivores eat plants, they incorporate nitrogen into their own tissues.

4. **Denitrification: Denitrifying bacteria** convert nitrate back into atmospheric nitrogen.

How do human activities affect the nitrogen cycle?

Fossil fuel combustion and burning organic matter release nitrogen into the atmosphere, where it forms acid rain. Chemical fertilizers also contain nitrogen, which escapes into the atmosphere or **leaches** into lakes and streams. High levels of nitrogen cause **eutrophication** (too many nutrients) and increased algal growth in aquatic ecosystems, depriving aquatic organisms of sunlight and oxygen.

How does the phosphorus cycle work?

Phosphorus carries energy to cells. It is found in phosphate (PO_4^{3-}) rock and sediments on the ocean floor. **Weathering**— through **chemical** or **physical** means—breaks down rock, releasing phosphate into the soil from longer-term stores. Organisms take up phosphorus. When they die, decomposers return phosphorus to the soil. Excess phosphorus settles on floors of lakes and oceans, eventually forming sedimentary rock. It remains trapped for millions of years until it is exposed through **geologic uplift** or mountain building.

How do human activities affect the phosphorus cycle?

Commercial fertilizers and phosphate-containing detergents enter waterways and contribute additional phosphate to the phosphorus cycle. Slash-and-burn forest clearance reduces phosphate levels, as phosphate in trees enters soil as ash. It leaches out of the soil and settles on lake and ocean bottoms, unavailable to organisms. ✔

✔ Reading Check

1. List a human activity that can cause changes to a nutrient cycle.

_____

_____

_____

Use with textbook pages 68–87.

Nutrient cycles

Answer the questions below.

1. Where are nutrients accumulated or stored for short or long periods?

   _____

2. Name a biotic process and an abiotic process that allow nutrients to flow in and out of stores.

   _____

   _____

3. Photosynthesis is an important process in which carbon and oxygen are cycled through ecosystems. Describe this process.

   _____

   _____

   _____

4. Cellular respiration is the process in which plants and animals make use of stored energy and release carbon dioxide back into the atmosphere. Describe this process.

   _____

   _____

5. How is decomposition related to the carbon cycle?

   _____

   _____

6. What is nitrogen fixation?

   _____

7. What is denitrification?

   _____

8. What is eutrophication?

   _____

Use with textbook pages 69–70, 86–87.

The cycling of nutrients in the biosphere

Use the general model of a nutrient cycle to answer the questions below.

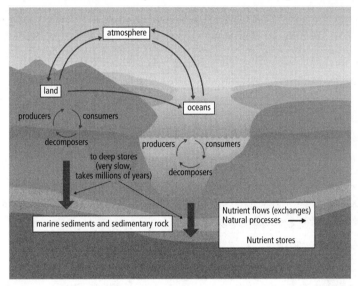

1. This diagram illustrates the general model of a nutrient cycle. What types of human activities can affect a nutrient cycle?

_____

_____

2. How do these human activities affect a nutrient cycle?

_____

_____

3. On the diagram above, add terms and arrows that could represent the effects of human activity on a nutrient cycle.

4. How do changes in nutrient cycles affect biodiversity?

_____

_____

5. Reflect on your local community. Discuss a human activity that is affecting your local ecosystem.

_____

_____

Use with textbook pages 71–87.

The carbon, nitrogen, and phosphorus cycles

The carbon cycle

Use the nutrient cycle below to answer the questions in the chart that follows.

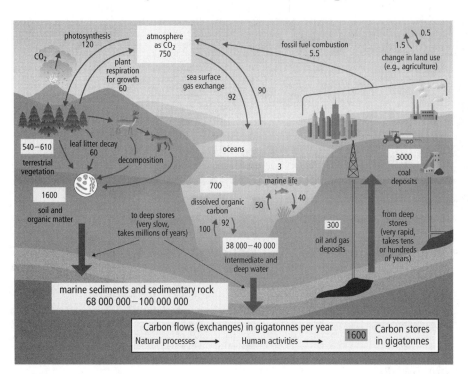

Why is the carbon cycle important?	
How is carbon stored?	
How is carbon cycled?	
Name several human activities that affect the carbon cycle.	

The nitrogen cycle

Use the nutrient cycle below to answer the questions that follow.

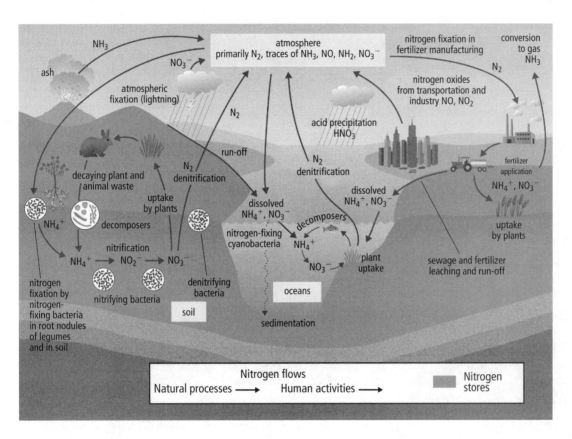

Why is the nitrogen cycle important?	
How is nitrogen stored?	
How is nitrogen cycled?	
Name several human activities that affect the nitrogen cycle.	

The phosphorus cycle

Use the nutrient cycle below to answer the questions that follow.

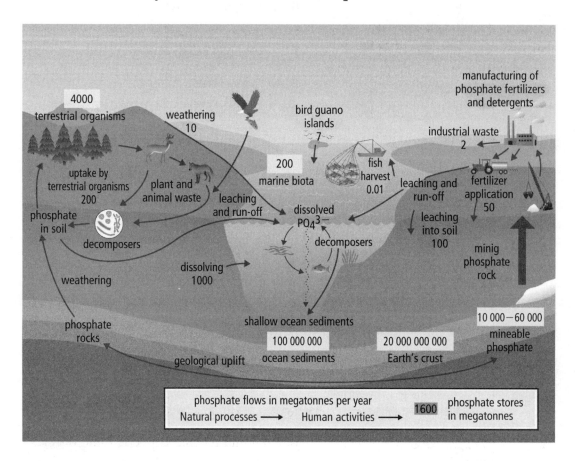

Why is the phosphorus cycle important?	
How is phosphorus stored?	
How is phosphorus cycled?	
Name several human activities that affect the phosphorus cycle.	

Nutrient cycles in ecosystems

Use with textbook pages 68–87.

Match each Term on the left with the best Descriptor on the right. Each Descriptor may be used only once.	
Term	**Descriptor**
1. _____ cellular respiration **2.** _____ denitrification **3.** _____ nitrification **4.** _____ nutrients **5.** _____ photosynthesis **6.** _____ sedimentation **7.** _____ weathering	**A.** the process in which nitrogen is released into the atmosphere **B.** substances, such as nitrogen and phosphorus, that are required by plants and animals for energy, growth, development, repair, and maintenance **C.** the process in which rock is broken into smaller fragments **D.** a process in which carbon dioxide enters plants and reacts with water in the presence of sunlight to produce carbohydrates and oxygen **E.** the process in which ammonium is converted into nitrate **F.** the process in which plants and animals release carbon dioxide back into the atmosphere by converting carbohydrates and oxygen into carbon dioxide and water. **G.** the process in which soil particles and decaying organic matter accumulate in layer on the ground or at the bottom of large bodies of water

Circle the letter of the best answer.

8. In the carbon cycle, where are the highest stores of carbon found?

  A. terrestrial vegetation

  B. marine sediments and sedimentary rocks

  C. oil and gas deposits

  D. soil and organic matter

9. Calcium carbonate is a structural component of:

  A. marine organisms

  B. terrestrial organisms

  C. algae

  D. volcanic ash

10. Which of the following is not stored in the atmosphere as a gas?

  A. carbon

  B. oxygen

  C. nitrogen

  D. phosphorus

11. Nitrogen fixation results in:

  A. ammonium being converted into nitrates

  B. nitrates being consumed by bacteria

  C. nitrogen gas being converted into nitrate or ammonium

  D. ammonia being converted into carbohydrates

12. Lightning provides energy that:

  A. absorbs energy into land masses

  B. fixes nitrogen in the atmosphere

  C. fixes carbon dioxide in the atmosphere

  D. releases nitrogen into the soil

Effects of Bioaccumulation on Ecosystems

Textbook pages 92–103

Before You Read

Everyday activities, such as driving or heating your home, often pollute ecosystems. In your opinion, which human activity is the most harmful to the environment? Explain.

_____

_____

 Mark the Text

Summarize
As you read this section, highlight the main points in each paragraph. Then write a short paragraph summarizing what you have learned.

How can pollutants affect food chains and ecosystems?

Human activity creates many harmful pollutants. These build up in the environment when decomposers are unable to break them down. Plants take up these pollutants. The pollutants are then transferred along the food chain until they reach the highest trophic level. **Bioaccumulation** refers to the gradual build-up of pollutants in living organisms. **Biomagnification** refers to the process in which pollutants not only accumulate, but also become more concentrated at each trophic level. Organisms at lower trophic levels may be affected by the pollutant, but primary, secondary, and tertiary consumers will be more affected, because levels will build up in their tissues as they consume contaminated food. An example of this is the PCB concentrations in the orca's food web. When orcas consume food contaminated with PCBs, they store some of the PCBs in their blubber. When salmon (their primary food) is not available, orcas use their blubber for energy. This releases PCBs into their system. Pollutants can build up to toxic levels in organisms at the top of the food chain. They can also affect entire ecosystems when **keystone species**, species that greatly affect ecosystem health, or the reproductive abilities of species are harmed. ✔

✔ **Reading Check**

1. What is the difference between bioaccumulation and biomagnification?

_____

_____

_____

_____

_____

_____

What are some human-made compounds that bioaccumulate and biomagnify?

PCBs (polychlorinated biphenyls)

◆ PCBs were once widely used in industrial products but are now banned in North America. They interfere with normal functioning of the body's immune system and cause problems with reproduction.

◆ PCBs have a long **half-life** (time it takes for the amount of a substance to decrease by half). They stay in the environment for a long time. Aquatic ecosystems are most sensitive to PCBs. Organisms at high trophic levels, like the orca, retain high levels of the pollutant.

POPs (persistent organic pollutants)

◆ POPs are harmful, carbon-containing compounds that remain in water and soil for many years.

◆ **DDT** (dichloro-diphenyl-trichloroethane) is a toxic POP that was used as a **pesticide** in the past to control disease-carrying mosquitoes.

◆ Accumulation is measured in **parts per million (ppm)**. This refers to one particle of a given substance mixed with 999 999 other particles. DDT is harmful at 5 ppm.

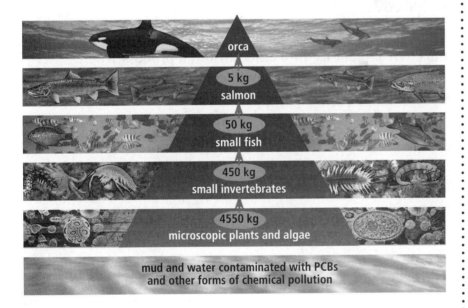

orca

5 kg
salmon

50 kg
small fish

450 kg
small invertebrates

4550 kg
microscopic plants and algae

mud and water contaminated with PCBs
and other forms of chemical pollution

Heavy metals

Once heavy metals enter the biosphere, they do not degrade, and they can not be destroyed.

◆ Heavy metals, such as lead (Pb), cadmium (Cd), and mercury (Hg), are toxic at low concentrations; however, small amounts are naturally present in soil. For humans, the most serious source of cadmium poisoning is smoking.

◆ Human activities can cause these metals to build up in ecosystems. In the past, use of lead-based insecticides, batteries, and paints, increased lead to harmful levels. Despite reductions, lead still enters ecosystems through improperly disposed electronic waste.

How can the effects of chemical pollution be reduced?

Some harmful chemical pollutants can be removed from the environment by **bioremediation,** a process where micro-organisms or plants help clean them up. Reacting contaminants with certain chemicals can also make them less harmful. ✔

Reading Check

1. Provide an example of how the effects of chemical pollution can be reduced.

_____

_____

_____

Use with textbook pages 92–99.

Bioaccumulation

Vocabulary	
bioaccumulation	lead
biomagnification	mercury
bioremediation	parts per million
cadmium	PCBs
half-life	persistent organic pollutants
heavy metals	producers
keystone species	

Use the terms in the vocabulary box to fill in the blanks. Use each term only once.

1. _____ is the gradual build-up of synthetic and organic chemicals in living organisms.

2. _____ are species that can greatly affect population numbers and the health of an ecosystem.

3. _____ is the process in which chemicals not only accumulate but become more concentrated at each tropic level in a food pyramid.

4. Even small concentrations of chemicals in _____ and primary and secondary consumers can build up to cause problems in higher trophic levels.

5. _____ are synthetic chemicals that were widely used from the 1930s to the 1970s in industrial products.

6. _____ is the time it takes for the amount of a chemical to decrease by half.

7. _____ are carbon-containing compounds that remain in water and soil for many years.

8. Chemical accumulation is measured in _____

9. _____ are metallic elements with a high density that are toxic to organisms at low concentrations.

10. Three polluting heavy metals are _____, _____, and _____.

11. _____ is the use of living organisms to clean up chemical pollution naturally, only faster, through biodegradation.

Use with textbook pages 94–98.

Impact of bioaccumulation on consumers

Complete the following table to demonstrate the effects of each of these chemicals
on various trophic levels in their ecosystems.

Chemical	Effects on producers, primary consumers, and secondary consumers	Effects on humans
toxic organic chemicals from red tide		
DDT		

lead 		
cadmium 		
mercury 		

Use with textbook page 95.

PCBs and the orca

1. What are PCBs? What is their full chemical name?

_____

2. What were PCBs used for in the 1970s?

_____

_____

3. In North America, PCBs were banned in 1977. Explain why they are still having an effect on organisms today.

_____

_____

_____

4. Explain what happens to PCBs when they enter an orca's body.

_____

_____

5. How do orcas survive when salmon stocks are low? What effect does this have on their survival?

_____

_____

_____

6. Draw a diagram to illustrate how biomagnification occurs in orcas.

Effects of bioaccumulation on ecosystems

Use with textbook pages 92–99.

Match each Term on the left with the best Descriptor on the right. Each Descriptor may be used only once.	
Term	**Descriptor**
1. _____ bioaccumulation **2.** _____ bioremediation **3.** _____ heavy metals **4.** _____ keystone species **5.** _____ parts per million **6.** _____ PCBs	**A.** synthetic chemicals containing chlorine that are used in the manufacture of plastics and other industrial products **B.** species that can greatly affect population numbers and the health of an ecosystem **C.** a measurement of chemical accumulation **D.** the use of organisms to break down chemical pollutants in water or soil to reverse or lessen environmental damage **E.** metallic elements with a high density that are toxic to organisms at low concentrations **F.** the gradual build-up of synthetic and organic chemicals in living organisms

Circle the letter of the best answer.

7. Over the last century, which human activity has caused the greatest change to the environment?

A. recycling

B. forest fires

C. introduction of synthetic chemicals

D. building of hydro plants

8. Which of the following would be identified as a keystone species in the BC forest ecosystem?

A. bacteria

B. fungi

C. pine trees

D. salmon

9. POPs, or persistent organic pollutants, are compounds that contain:

A. oxygen

B. carbon

C. phosphorus

D. nitrogen

10. For humans, the most serious source of cadmium poisoning is exposure to:

A. air pollution

B. water pollution

C. tobacco smoke

D. pesticides

11. Within the biosphere, heavy metals:

A. do not degrade and cannot be destroyed

B. do not degrade and can be destroyed

C. do degrade and can be recycled

D. do degrade and can not be recycled

12. The process by which microorganisms break down chemical pollutants to lessen environmental damage is known as:

A. bioaccumulation

B. biodiversity

C. biomagnification

D. bioremediation

How Changes Occur Naturally in Ecosystems

Textbook pages 108–121

Before You Read

How do you think mature forests, such as the temperate rainforests of coastal British Columbia, change over time? Write your answer on the lines below.

_____

_____

 ? *Create an Outline*

Create an outline highlighting the typical changes that occur in an ecosystem undergoing primary succession.

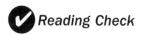 *Reading Check*

1. What is adaptive radiation?

_____

_____

_____

_____

How do organisms adapt to change?

In **natural selection,** the best-adapted members of a species survive to reproduce. These individuals may pass favourable characteristics on to their offspring. As abiotic and biotic components of their environment change, **adaptive radiation** may result. This term describes the change from a common ancestor into a number of different species that "radiate out" to inhabit different niches. For example, 13 species of finches that fill different niches on the Galapagos Islands are thought to have developed from a single species from mainland South America. ✓

How do ecosystems change over time?

Ecological succession refers to changes that take place over time in the types of organisms that live in an area. There are two types of ecological succession:

1. Primary succession: **Primary succession** occurs in areas where no soil exists, such as following glaciation or a lava flow. Wind and rain carry spores of lichens to these areas. Lichens obtain nutrients by secreting chemicals that break down rock. As lichens decay, they add organic matter to the developing soil. The first organisms to survive and reproduce in an area are called **pioneer species.** They are adapted to grow in harsh, nutrient-poor conditions. In time, often over hundreds of years, the weathering of rocks and decay of pioneer species cause soil formation. The abiotic conditions of the ecosystem continue to change as new species of plants and animals colonize the area, each competing for nutrients, moisture, and sunlight. More niches are created and biodiversity increases.

Eventually, primary succession leads to the development of mature **climax communities,** such as a boreal forest or grassland.

2. Secondary succession: Small disturbances, such a fire, often occur in ecosystems. **Secondary succession—** succession that occurs as a result of a disturbance to an area that already has soil and was once the home of living organisms—occurs as a result. It proceeds much faster than primary succession since micro-organisms, insects, seeds, and nutrients still exist in the soil. ✔

How do natural events affect ecosystems?

Natural events can destroy habitats, reduce biodiversity, and cause regions to undergo succession. Some examples include:

- flooding: results in soil erosion, pollution, and disease when toxins or harmful bacteria from untreated sewage enter drinking water
- drought: plants and animals die due to lack of water
- insect infestations: often result in succession in forests because insects destroy older, weaker trees
- tsunamis: huge, rapidly moving ocean waves destroy habitats and salt water carried onto shore changes soil composition

✔ *Reading Check*

2. Describe the difference between primary and secondary succession.

_____

_____

_____

_____

Use with textbook pages 108–117.

Change in ecosystems

Vocabulary	
adaptive radiation climax community drought ecological succession flooding insect infestations	natural selection pioneer species primary succession secondary succession tsunami

Use terms in the vocabulary box to fill in the blanks. Use each term only once.

1. In the process of _____ , living organisms change as the abiotic and biotic components in their environment change.

2. _____ describes the change from a common ancestor into a number of different species that "radiate out" to inhabit different niches.

3. Scientists use the term _____ to refer to changes that take place over time in the types of organisms that live in an area.

4. _____ occurs in an area where no soil exists, such as on bare rock.

5. The lichens and others plants that are the first organisms to survive and reproduce in an area are known as _____ .

6. The process of primary succession leads to the development of a mature community, which is sometimes called a _____ .

7. _____ occurs as the result of a disturbance to an area that already has soil and was once the home of living organisms.

8. _____ can result in soil erosion and soil pollution if toxic chemicals are present in floodwaters.

9. _____ is a huge, rapidly moving ocean wave.

10. _____ can result in crop failures and livestock deaths.

11. _____ , such as the mountain pine beetle in the forests of British Columbia, have a devastating effect on the forest canopy, and bird and mammal habitats.

Use with textbook pages 111–114.

Primary and secondary succession

1.

Glacier Retreating

As a glacier retreats, the process of primary succession will occur. Describe the various stages that lead to the development of a mature community.

_____

_____

_____

_____

_____

2.

Results of Forest Fire

After a forest fire, not much is left except ash and burnt trees. Describe the sequence of events that will occur during secondary succession.

_____

_____

_____

_____

_____

Use with textbook pages 115–117.

How natural events affect ecosystems

For each major event listed below, summarize the effects on their mature communities.

Natural event	Effects on mature community
Fire	
Flooding	
Tsunami	
Drought	
Insect infestation	

Use with textbook pages 108–117.

How changes occur naturally in ecosystems

Match each Term on the left with the best Descriptor on the right. Each Descriptor may be used only once.	
Term	**Descriptor**
1. _____ adaptive radiation **2.** _____ climax community **3.** _____ ecological succession **4.** _____ natural selection **5.** _____ pioneer species	**A.** a mature community that continues to change over time **B.** the development of a number of new species from a common ancestor **C.** organisms, such as lichens and other plants, that are the first to survive and reproduce in an area **D.** changes that take place over time in the types of organisms that live in an area **E.** the process in which, over time, the best adapted members of a species will survive and reproduce

Circle the letter of the best answer.

6. The process that makes change possible in living things is

A. ecological succession

B. primary succession

C. natural selection

D. adaptive radiation

7. Each of these finches from the Galapagos Islands has evolved different shapes and sizes of beaks. This is an example of:

A. primary succession

B. secondary succession

C. bioremediation

D. adaptive radiation

0 Years ——————————————→ 300 Years

8. The diagram above represents which of the following:

A. adaptive radiation

B. climax community

C. ecological succession

D. natural succession

9. An example of a pioneer species would be

A. moss

B. lichen

C. deciduous trees

D. coniferous trees

How Humans Influence Ecosystems

Textbook pages 122–137

Before You Read

What do you think of when you hear or read the term sustainability? What does this term refer to? Record your thoughts on the lines below.

Mark the Text

In Your Own Words

Write a paragraph explaining three ways resource exploitation may harm ecosystems.

✔ **Reading Check**

What is sustainability?

_____

_____

_____

_____

_____

✔ **Reading Check**

List two ways overexploitation can affect organisms.

_____

_____

_____

_____

How does land and resource use affect sustainability?

Sustainability refers to the ability of an ecosystem to sustain ecological processes. We make many demands on nature through our use of land and resources. **Land use** refers to the ways we use the land around us—for cities, roads, industry, agriculture, and recreation. **Resource use** refers to the ways we obtain and use resources—naturally occurring materials, such as soil, wood, water, and minerals. Resource use is also referred to as **resource exploitation.** One example of resource use affecting sustainability is deforestation in China. The result is that less bamboo is available as food for China's giant pandas. In another example, whaling in the Pacific Ocean decreased numbers of whales—the orcas' primary food source. Orcas turned to eating other prey, such as sea otters. But sea otters eat sea urchins and when the numbers of sea otters went down, the numbers of sea urchins exploded.

How can First Nations' knowledge improve resource management?

First Nations' thorough understanding of the plants, animals, and natural occurrences in their environment is referred to as **traditional ecological knowledge.** It reflects knowledge—about local climate and resources, biotic and abiotic characteristics, and animal and plant life cycles—that was gained over centuries. It provides researchers with valuable data with regard to management practices that enhance the productivity and health of ecosystems.

How can resource exploitation affect ecosystems?

Certain effects of resource exploitation, such as those described in the table below, can affect the biodiversity and health of ecosystems.

Effect	Example of human activity	How ecosystems are affected
habitat loss	Humans take over natural space in the creation of cities and agriculture.	Habitats are destroyed and can no longer support the species that lived there.
habitat fragmentation	Agriculture, roads, and cities divide natural ecosystems into smaller, isolated fragments.	Plant pollination, seed dispersal, wildlife movement, and reproduction are adversely affected.
deforestation	Forests are logged or cleared for human use and never replanted.	The number of plants and animals living in an ecosystem are reduced.
soil degradation	Deforestation and land clearance leave land bare so water and wind erosion remove topsoil.	Organic matter, water, and nutrients are removed along with the topsoil, reducing plant growth.
soil compaction	Agricultural farm vehicles and grazing animals squeeze soil particles together.	Reduces the movement of air, water, and soil organisms in soil, hindering the growth of plants and increasing run-off of fertilizer and pesticides.
contamination	By-products of resource exploitation, such as mining, introduce toxins.	Toxins are introduced into the environment in harmful concentrations and kill plants and animals.
overexploitation	A resource—like fish or forests—is used or extracted until it is depleted.	Food web interactions are affected. Organisms become less resistant to disease and less able to adapt to environmental change. Extinction, the dying out of a species, can result. ✓

Use with textbook pages 125–134.

Sustainability

1. What is sustainability?

_____

_____

2. What is the difference between the terms habitat loss and habitat fragmentation?

_____

_____

3. What is deforestation? What are the consequences of deforestation?

_____

_____

_____

4. What are the advantages of aeration, or breaking up compacted soil?

_____

_____

5. List four examples of contamination that can occur due to mining.

_____

_____

6. Explain the effects on an ecosystem when resources are overexploited.

_____

_____

_____

7. Define the term traditional ecological knowledge. Summarize the various factors taken into consideration when traditional ecological knowledge is used to examine an ecosystem.

_____

_____

_____

_____

_____

Use with textbook pages 126–134.

Effects of human activities on ecosystems

Summarize the possible effects on an ecosystem due to each of the following human activities.

Human activity	Effects on ecosystem
deforestation	
agricultural practices, such as leaving fields bare during non-planting seasons	
exploitation of mining resources	
overexploitation of natural resources, such as fish	

Use with textbook pages 125–134.

Sustainability

In British Columbia, we often use land in ways that affect natural ecosystems, such as switching to agriculture, expanding our urban areas, and cutting down trees. For each of the following examples, describe the effects on the habitat and suggest a sustainable approach that could help the local ecosystem survive.

Example of land use	Effects on habitat	Sustainable approach suggestions
the conversion of grasslands into cattle ranches in the Interior of British Columbia		
clear-cutting of forests on Vancouver Island		
urbanization of the Fraser Valley		

Use with textbook pages 122–134.

How humans influence ecosystems

Match each Term on the left with the best Descriptor on the right. Each Descriptor may be used only once.	
Term	**Descriptor**
1. _____ deforestation 2. _____ extinction 3. _____ habitat loss 4. _____ soil compaction 5. _____ soil degradation 6. _____ sustainability 7. _____ traditional ecological knowledge	**A.** the ability of an eco-system to sustain eco-logical processes and maintain biodiversity over time **B.** the clearing or log-ging of forests without replanting **C.** ecological information passed down from generation to gen-eration, which reflects human experience with nature **D.** the dying out of a species **E.** the squeezing together of soil particles so that the air spaces between them are reduced **F.** damage to soil **G.** the destruction of habitats that usually results from human activities

Circle the letter of the best answer.

8. Which of the following illustrates a sustainable practice?

   A. conversion of grassland into ranchland

   B. urban expansion of cities

   C. restoration of a streambeds

   D. extraction of gold in mining

9. Which of the following factors has lead to the giant panda in China being considered an endangered species?

   A. soil degradation

   B. overexploitation

   C. contamination of ecosystem

   D. deforestation

10. In the Pacific Ocean, the food web, including kelp, whales, sea otters, and sea urchins, has been altered by human activities. What factor has been linked to the explosion in the sea urchin population?

    A. decrease in the sea otter population

    B. increase in kelp beds

    C. change in migration pattern of orcas

    D. increase in fur seal population

11. Which of the following is an example of traditional ecological knowledge practices?

    A. habitat fragmentation by urbanization

    B. grassland management by controlled burning

    C. resource exploitation by mining industry

    D. clear-cutting practices by forestry industry

How Introduced Species Affect Ecosystems

Textbook pages 138–147

Before You Read

Invasive species can dramatically change or destroy ecosystems. Do you think unwanted weeds, such as dandelions, are invasive species? Record your thoughts on the lines below.

_____

_____

 Create an Outline

Create a quiz to help you learn the concepts introduced in this section. Answer your questions and share your quiz with your classmates.

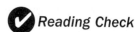 **Reading Check**

Are all introduced species invasive species?

_____

_____

_____

_____

Reading Check

Describe a way introduced invasive species may make natural habitats unsuitable for native species.

_____

_____

_____

_____

How can introduced species affect an ecosystem?

Native species are plants and animals that naturally inhabit an area. **Introduced species** or **foreign species** are species that have been introduced into an ecosystem by humans, either intentionally or accidentally. They do not naturally inhabit the ecosystem. Introduced species are usually beneficial or harmless. However, some introduced species, known as **invasive species**, can dramatically change or destroy ecosystems. With climate change and the expansion of international trade and travel, invasive species are entering new ecosystems at an increasing rate. This rapid spread of introduced invasive species is a major cause of global biodiversity loss. Introduced species can affect native species through competition, predation, disease, parasitism, and habitat alteration, as shown in the table on the next page. ✔

Examples of the effects of introduced species include:

◆ Scotch broom was introduced to British Columbia as a garden plant. It has up to 18 000 seeds per plant, can survive drought, and fixes nitrogen in the soil, causing conditions that many native species have trouble growing in. Together with other introduced species, it is competing with the keystone species Garry oak on Vancouver Island.

◆ European starlings outcompete native birds for nesting sites, and cause decreases in their populations. Barn owls are able to keep the numbers of starlings low in some areas.

- Eurasian milfoil forms mats on the surface of waterways that decrease the amount of sunlight available to organisms lower down. It is spread by boat traffic since it can regrow from small pieces.
- Norway rats eat a wide variety of foods, including preying on native species.
- Blister rust grows on the native white whitebark pine, causing disease that kills the trees.
- Wild boars are considered one of the world's worst invasive species. Their behaviours spread weeds and interfere with natural succession, and they prey on native species.
- Purple loosetrife is an introduced species, and the introduced loosestrife-eating beetle is proving to be a good way to control its spread.

Effect	Harm to native species	Example
competition	Introduced invasive species reproduce rapidly and are often aggressive. Lacking natural predators, they easily outcompete native species for food and habitat.	Invasive carpet burweed from South America competes with rare native plants for habitat. Its spiny tips protect it from predators.
predation	Introduced predators can have more impact on a prey population than native predators, as prey may not have adaptations to escape or fight them.	Yellow crazy ants that escaped from cargo from West Africa have devastated the population of red crabs, a keystone species on Christmas Island, Australia.
disease and parasites	An invasion of parasites or disease-causing viruses and bacteria can weaken the immune responses of native plants and animals.	The sea lamprey, a parasitic fish that has invaded freshwater ecosystems in eastern Canada, sucks body fluids from its prey by attaching to it with a sucker-like mouth.
habitat alteration	Introduced invasive species can make a natural habitat unsuitable for native species by changing its structure or composition.	Wild boars damage ecosystems by rooting, wallowing, and spreading weeds that interfere with natural succession. ✅

Use with textbook pages 138–144.

Introduced species

1. What are native species?

_____

2. What is an invasive species?

_____

3. What impact can an invasive species have on an ecosystem?

_____

_____

4. List five ways an introduced species can affect a native species.

_____

_____

5. List three examples of introduced invasive species that have had an impact on ecosystems in British Columbia.

_____

6. On Vancouver Island, three invasive species have dominated the plant cover. Which keystone species are they competing with?

_____

7. Why is Scotch broom such a tough species to control?

_____

_____

_____

Use with textbook pages 139–142.

The impact of introduced invasive species

Give an example of how an invasive species has used each of the following methods to invade an ecosystem or destroy a species.

Method	Invasive species	Effect on ecosystem
competition		
predation		
disease and/or parasites		
habitat alteration		

Use with textbook pages 138–144.

Invasive species in British Columbia

In British Columbia, many introduced invasive species are having an impact on ecosystems. Species such as purple loosestrife, Eurasian milfoil, Norway rat, American bullfrog, European starling, and Scotch broom are just some of the invasive species affecting British Columbia's ecosystems.

Select three of the above introduced invasive species. Draw a picture of the species. Explain how they were introduced and what effect they are having on the ecosystem.

Species	Method of introduction	Effect on environment

Use with textbook pages 138–144.

How introduced species affect ecosystems

Match each Term on the left with the best Descriptor on the right. Each Descriptor may be used only once.	
Term	**Descriptor**
1. _____ biodiversity 2. _____ competition 3. _____ introduced species 4. _____ invasive species 5. _____ keystone species 6. _____ native species 7. _____ predation	**A.** a harmful interaction between two or more organisms that can occur when organisms compete for the same resource **B.** species that greatly affect population numbers and the health of an ecosystem **C.** predator-prey interactions in which one organism eats all or part of another organism **D.** introduced organisms that can take over the habitats of native species or invade their bodies **E.** the variety of all living species of plants, animals, and micro-organisms on Earth **F.** plants and animals that naturally inhabit an area **G.** plants, animals, or micro-organisms that are transported intentionally or by accident into regions in which they did not exist previously

Circle the letter of the best answer.

8. Which of the following introduced species has been found not to be harmful to its environment?

    A. loosestrife eating beetle

    B. blister rust

    C. zebra mussels

    D. lamprey

9. Why do introduced predators have more of an impact on a prey population than native predators?

    A. prey lack adaptations to escape

    B. prey populations increase

    C. prey adapt to new predators

    D. native predators' populations decrease

10. Which animal is considered the world's most invasive species?

    A. yellow stinging ant

    B. maggot

    C. rattle snake

    D. wild boar

11. In some agricultural areas of British Columbia, how are the populations of European starlings being controlled?

    A. limiting planting of grain crops

    B. introducing barn owls

    C. removing European starlings' nests

    D. releasing competitive species

12. An invader of the Garry Oak ecosystem is:

    A. red squirrel

    B. Norway rat

    C. Scotch broom

    D. American bullfrog

Atomic Theory and Bonding

Textbook pages 168–183

Before You Read

What do you already know about Bohr diagrams? Record your answer in the lines below.

_____

_____

What are atoms?

An **atom** is the smallest particle of any element that retains the properties of the element.

The particles that make up an atom are called **subatomic particles**. Atoms are composed of three subatomic particles: protons, neutrons, and electrons.

Name	Symbol	Electric Charge	Location in the Atom	Relative Mass
Proton	p	1+	Nucleus	1836
Neutron	n	0	Nucleus	1837
Electron	e	1−	Surrounding the nucleus	1

Nuclear charge is the electric charge on the nucleus. This charge is always positive, since the protons have a positive charge and the neutrons are not charged. **Atomic number** is the number of protons. The nuclear charge or atomic number is given in the top left hand corner of the element box for each element in the periodic table.

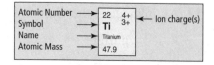

How does the periodic table provide information about elements?

In the periodic table, each element is listed according to its atomic number. Each row is called a **period**. Each column

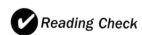

Mark the Text

Identify Definitions

Highlight the definition of each word that appears in bold type.

✔ **Reading Check**

Which has a positive electric charge, a proton, a neutron, or an electron?

is called a **group** or **family**. Metals are on the left side and in the middle of the table. Non-metals are in the upper right corner. The metalloids form a staircase toward the right side. The block of elements from groups 3 through 12 are the transition metals. Elements in the same chemical group or family have similar chemical properties. For example, group 17 contains very reactive non-metals known as halogens (i.e., fluorine, chlorine, bromine, etc.) Group 18 contains the non-reactive noble gases.

How do Bohr diagrams represent atoms?

A **Bohr diagram** shows the arrangement of subatomic particles in atoms and ions. Electrons are organized in "shells". The first shell holds a maximum of two electrons; the second shell a maximum of eight. When this shell is filled, it is called a **stable octet**. The outermost shell containing electrons is called the **valence shell**. The electrons in this shell are called **valence electrons**. These electrons are involved in chemical bonding. When an atom forms a compound, it acquires a full valence shell of electrons and achieves a stable, low energy state. On the periodic table, elements in Group 1 have 1 electron in their valence shell, elements in Group 2 have 2 (a **lone pair**), elements in Group 3 have 3, and so on.

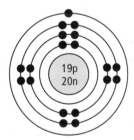

The Bohr diagram for a potassium atom

What are ionic and covalent compounds?

There are two basic types of compounds: ionic and covalent.

1. Ionic compounds: When atoms gain or lose electrons, they become electrically charged particles called **ions**. An ionic compound contains a positive ion (usually a metal) and a negative ion (usually a non-metal). In **ionic bonding**, one or more electrons transfer from each atom of the metal

to each atom of the non-metal. The metal atoms lose electrons, forming **cations**. For example, aluminum forms a 3+ cation as a result of losing three electrons. Some metals are **multivalent** and can form ions in several ways, depending on the chemical reaction they undergo. For example, iron is multivalent because it can lose two or three electrons to become a Fe^{2+} or Fe^{3+} ion. The non-metal atoms gain electrons, forming **anions**. Chlorine gains one electron and forms a 1^- anion.

The common ions are sometimes shown in the upper right-hand corner of the element's box in the periodic table. For a multivalent metal, the most common charge is listed first.

2. Covalent compounds: In **covalent bonding**, the atoms of a non-metal share electrons with other non-metal atoms. An unpaired electron from each atom will pair together, forming a **covalent bond**. These two electrons are sometimes called a **bonding pair**.

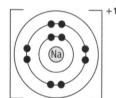 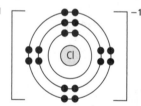

The ionic compound sodium chloride

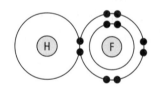

The covalent compound hydrogen fluoride

✔ **Reading Check**

What is a Lewis diagram?

_____

_____

_____

_____

What is a Lewis diagram?

A **Lewis diagram** illustrates chemical bonding by showing only an atom's valence electrons and its chemical symbol. Lewis diagrams can be used to represent elements, ions, and compounds. ✔

H·

hydrogen atom

$[:\ddot{C}l:]^-$

chlorine ion

$[Mg]^{2+}[:\ddot{O}:]^{2-}$

magnesium oxide molecule

Periodic Table of the Elements

Based on mass of C-12 at 12.00.

Any value in parentheses is the mass of the most stable or best known isotope for elements that do not occur naturally.

Use with textbook pages 168–180.

The atom and the subatomic particles

1. Use the following vocabulary words to label the diagram.

Vocabulary
common ion charge   symbol other ion charge   atomic number name   average atomic mass

(a) _____

(b) _____

(c) _____

(d) _____

22   4+
Ti   3+
Titanium
47.9

(e) _____

(f) _____

2. Examine the periodic table for the element below and complete the blanks.

35   –
Br
Bromine
79.9

(a) atomic number _____

(b) average atomic mass _____

(c) ion charge _____

(d) number of protons _____

(e) name of element _____

(f) number of neutrons _____

3. Complete the following table for the different atoms and ions. The first two rows have been completed to help you.

Element Name	Atomic Number	Ion Charge	Number of Protons	Number of Electrons	Number of Neutrons
potassium	19	1+	19	18	20
phosphorus	15	0	15	15	16
	3	0			
		2+	20		
nitrogen		3–			
	5	0			
argon				18	
	13			10	
chlorine		0			
			11	10	

Use with textbook pages 174–177.

Bohr diagrams

1. Define the following terms:

(a) Bohr diagram _____

(b) stable octet _____

(c) valence shell _____

(d) valence electrons _____

2. Complete the following table.

Atom/Ion	Atomic Number	Number of Protons	Number of Electrons	Number of Neutrons	Number of Electron Shells
neon atom					
fluorine atom					
fluoride ion					
sodium atom					
sodium ion					

3. Use the table above to draw the Bohr model diagram for each of the following atoms and ions.

neon atom	fluorine atom	fluoride ion	sodium atom	sodium ion

4. Draw the Bohr model diagram for each of the following compounds.

carbon dioxide (CO_2)	ammonia (NH_3)	calcium chloride ($CaCl_2$)

Use with textbook pages 176–180.

Lewis diagrams

1. Define the following terms:

(a) Lewis diagram

(b) lone pair _____

(c) bonding pair _____

2. Draw Lewis diagrams for each of the following elements.

(a) boron   (b) nitrogen   (c) aluminium   (d) chlorine

3. Draw Lewis diagrams for each of the following ionic compounds.

(a) sodium oxide   (b) potassium chloride   (c) magnesium bromide

4. Draw Lewis diagrams for each of the following covalent compounds.

(a) carbon dioxide, CO_2 (b) phosphorus trifluoride, PF_3 (c) silicon tetrachloride, $SiCl_4$

5. Draw Lewis diagrams for each of the following diatomic molecules.

(a) chlorine, Cl_2 (b) nitrogen, N_2 (c) hydrogen, H_2

Use with textbook pages 168–180.

Atomic theory and bonding

Match the Term on the left with the best Descriptor on the right. Each Descriptor may be used only once.	
Term	**Descriptor**
1. _____ shell **2.** _____ period **3.** _____ family **4.** _____ ionic bonding **5.** _____ covalent bonding	**A.** a horizontal row on the periodic table **B.** a vertical column on the periodic table **C.** an area around the nucleus where electrons exist **D.** chemical bonding that results from a sharing of valence electrons **E.** chemical bonding that results when one or more electrons transfers from each atom of a metal to each atom of a non-metal

6. Which of the following is the smallest particle of an element that can exist by itself?

A. ion

B. atom

C. molecule

D. compound

7. Which of the following correctly matches the subatomic particle with its charge and location in an atom?

	Subatomic Particle	Location	Charge
A.	proton	nucleus	neutral
B.	neutron	nucleus	positive
C.	electron	shell	positive
D.	electron	shell	negative

8. Which of the following are responsible for bonding?

A. nuclei

B. protons

C. neutrons

D. electrons

Use the following diagram of an atom to answer questions 9 to 12.

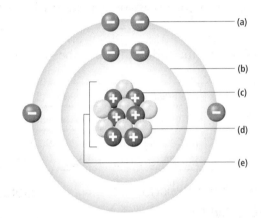

9. Which labelled part in the diagram represents a neutron?

A. (a)

B. (b)

C. (c)

D. (d)

10. What is the number of subatomic particle (c) equivalent to?

A. atomic number

B. mass number – atomic number

C. mass number + atomic number

D. number of electrons + number of protons

11. How many valence electrons are there in this atom?

A. 2

B. 4

C. 6

D. 7

12. Which of the following describes structure (e)?

	CHARGE	SUBATOMIC PARTICLE(S) PRESENT
A.	neutral	electrons and neutrons
B.	positive	protons and neutrons
C.	positive	protons and electrons
D.	negative	electrons

13. Which of the following describes a cation?

I.	examples include Ca^{2+} and Al^{3+}
II.	a metal atom that has lost electrons
III.	has equal numbers of electrons and protons

A. I and II only

B. I and III only

C. II and III only

D. I, II, and III

14. Which row of the table is completed correctly for an atom of potassium?

	Atomic Number	Mass Number	Number of Protons	Number of Neutrons	Number of Electrons
A.	19	39	19	20	19
B.	19	39	39	20	20
C.	19	39	20	20	19
D.	39	19	19	19	20

Use the following Lewis diagrams of four hypothetical elements to answer question 15.

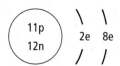

15. Which of the hypothetical elements shown above represents a metal?

A. Ma

B. Di

C. So

D. Nh

Use the following Bohr model of an element to answer question 16.

11p
12n
2e   8e

16. Which of the following does the Bohr model represent?

A. a neon atom

B. a sodium ion

C. a sodium atom

D. a fluorine atom

Names and Formulas of Compounds

Textbook pages 184–201

Before You Read

In this section, you will learn how to write the names and formulas of ionic and covalent compounds. Write what you already know about these compounds in the lines below.

_____

_____

How do you represent an ionic compound?

Ionic compounds are composed of positive and negative ions. They can be represented with both a name and a chemical formula.

1. Name: In an **ionic compound**, the first part of the name indicates the positive ion (a metal) and the second part indicates the negative ion (a non-metal). The non-metal's name always ends with the suffix "-ide." For example, lead sulphide.

2. Chemical formula: Follow the steps in the table below to write the chemical formula for an ionic compound.

Steps	Example Ionic Compound: Lead Sulphide
Identify the chemical symbol for each ion and its charge.	lead: Pb^{4+} sulphide: S^{2-}
Determine the total charges needed to balance the positive and negative charges of each ion.	Pb^{4+}: $+4 = +4$ S^{2-}: $-2\ -2 = -4$
Note the ratio of positive to negative ions.	$1\ Pb^{4+}$: $2\ S^{2-}$
Use these subscripts to write the chemical formula. Make sure the subscripts represent the smallest whole number formula. A "1" is not shown as a subscript.	PbS_2

There are also two special cases you must consider when naming and writing the chemical formulas of ionic compounds. These are compounds containing multivalent metals and polyatomic ions.

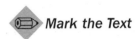
Mark the Text

Check for Understanding

As you read this section, be sure to reread any parts you do not understand. Highlight any sentences that help make concepts clearer for you.

1. **Multivalent metals:** Multivalent metals can form two or more positive ions with different ionic charges. To distinguish between two ions formed from multivalent metals, the name must contain the ion's charge. The Roman numerals I, II, III, IV, V, VI, and VII, corresponding to ion charges 1+ to 7+, are used for this purpose. The Roman numerals are included in the name of the compound. For example, nickel (II) chloride has the formula $NiCl_2$. Thus, nickel (II) has an ion charge of 2+. Nickel (III) has the formula $NiCl_3$. The ion charge of nickel (III) is 3+. ✔

2. **Polyatomic ions:** A **polyatomic ion** is an ion composed of more than one type of atom joined by covalent bonds. For example, carbonate (CO_3^{2-}) is a polyatomic atom. All polyatomic atoms have special names assigned to them. You will need to look these up in the following table when naming a compound that includes a polyatomic ion.

✔**Reading Check**

What is a multivalent metal?

_____

_____

_____

_____

Table 4.11 Names, Formulas, and Charges of Some Polyatomic Ions

Positive Ions		Negative Ions		
NH_4^+ ammonium	CH_3COO^- acetate	HCO_3^- hydrogen carbonate, bicarbonate	NO_2^- nitrite	
	CO_3^{2-} carbonate	HSO_4^- hydrogen sulfate, bisulfate	ClO_4^- perchlorate	
	ClO_3^- chlorate	HS^- hydrogen sulfide, bisulfide	MnO_4^- permanganate	
	ClO_2^- chlorite	HSO_3^- hydrogen sulfite, bisulfite	PO_4^{3-} phosphate	
	CrO_4^{2-} chromate	OH^- hydroxide	PO_3^{3-} phosphite	
	CN^- cyanide	ClO^- hypochlorite	SO_4^{2-} sulfate	
	$Cr_2O_7^{2-}$ dichromate	NO_3^- nitrate	SO_3^{2-} sulfite	

How do you represent a binary covalent compound?

A **binary covalent compound** contains two non-metal elements joined together by one or more covalent bonds. Like ionic compounds, binary covalent compounds can be represented with both a name and a chemical formula.

1. Name: When naming a binary covalent compound, prefixes are used to indicate how many atoms of each element are present. The second element's name ends with the suffix "-ide." For example, dinitrogen trioxide has two atoms of nitrogen and three atoms of oxygen. No prefix is used if there is just one atom of the first element. For example, carbon dioxide. The table below provides the first ten prefixes used to name binary covalent compounds. ✔

Prefix	Number of atoms
mono-	1
di-	2
tri-	3
tetra-	4
penta-	5
hexa-	6
hepta-	7
octa-	8
nona-	9
deca-	10

2. Chemical formula: When writing the chemical formula, subscripts are used to indicate the number of atoms present. For example, dinitrogen trioxide has the chemical formula N_2O_3. The exact number of atoms is always shown in the formula. For example, hydrogen peroxide is written as H_2O_2, not HO. Unlike the formula for an ionic compound, the subscripts do not always represent the smallest whole number formula.

✔ Reading Check

A certain element has 5 atoms in a binary covalent compound. Which prefix is used to name this element?

_____

Use with textbook pages 189–193.

Multivalent metals and polyatomic ions

1. Define the following terms:

(a) ionic compound

_____

(b) multivalent metal

_____

(c) polyatomic ion

_____

2. Write the formulae and names of the compounds with the following combination of ions. The first row is completed to help guide you.

	Positive ion	Negative ion	Formula	Compound name
(a)	Pb^{2+}	O^{2-}	PbO	lead(II) oxide
(b)	Sb^{3+}	S^{2-}		
(c)			TlCl	
(d)				tin(II) fluoride
(e)			Mo_2S_3	
(f)	Rh^{4+}	Br^-		
(g)				copper(I) telluride
(h)			NbI_5	
(i)	Pd^{2+}	Cl^-		

3. Write the chemical formula for each of the following compounds.

(a) manganese(II) chloride _____	(f) vanadium(V) oxide _____
(b) chromium(III) sulphide _____	(g) rhenium(VII) arsenide _____
(c) titanium(IV) oxide _____	(h) platinum(IV) nitride _____
(d) uranium(VI) fluoride _____	(i) nickel(II) cyanide _____
(e) nickel(II) sulphide _____	(j) bismuth(V) phosphide _____

4. Write the formulae for the compounds formed from the following ions. Then name the compounds.

	Ions		Formula	Compound name
(a)	K^+	NO_3^-	KNO_3	potassium nitrate
(b)	Ca^{2+}	CO_3^{2-}		
(c)	Li^+	HSO_4^-		
(d)	Mg^{2+}	SO_3^{2-}		
(e)	Sr^{2+}	CH_3COO^-		
(f)	NH_4^+	$Cr_2O_7^{2-}$		
(g)	Na^+	MnO_4^-		
(h)	Ag^+	ClO_3^-		
(i)	Cs^+	OH^-		
(j)	Ba^{2+}	CrO_4^{2-}		

5. Write the chemical formula for each of the following compounds.

(a) barium bisulphate _____	(f) calcium phosphate _____
(b) sodium chlorate _____	(g) aluminum sulphate _____
(c) potassium chromate _____	(h) cadmium carbonate _____
(d) calcium cyanide _____	(i) silver nitrite _____
(e) potassium hydroxide _____	(j) ammonium hydrogen carbonate _____

Name Date

Comprehension
Section 4.2

Use with textbook pages 186–196.

Chemical names and formulas of ionic compounds

1. Write the name for each of the following compounds.

(a) BeS

(b) Hg_3N_2

(c) $Cu(NO_3)_2$

(d) Ag_2O

(e) $CoBr_2$

(f) $Bi_3(PO_4)_5$

(g) CaF_2

(h) Mn_2O_3

(i) $Cr_2(SO_4)_3$

(j) $ZnCl_2$

(k) $Ni(OH)_2$

(l) $K_2Cr_2O_7$

(m) ScF_3

(n) NaI

(o) $Pb(CO_3)_2$

(p) $RbClO_2$

(q) K_3P

(r) $Mg(CN)_2$

(s) SnS

(t) $NaHCO_3$

2. Write the chemical formula for each of the following compounds.

(a) aluminum bromide

(b) platinum(II) sulphide

(c) strontium sulfite

(d) scandium oxide

(e) titanium(IV) nitrite

(f) ammonium sulphate

(g) lithium selenide

(h) lead(II) hydrogen sulphate

(i) sodium acetate

(j) cesium chloride

(k) cadmium(II) hydroxide

(l) zinc phosphate

(m) barium chloride

(n) tin(II) permanganate

(o) lithium hypochlorite

(p) gold(III) sulphate

(q) sodium nitrate

(r) chromium(III) chloride

(s) potassium carbonate

(t) iron(III) bisulphate

Name Date

Comprehension
Section 4.2

Use with textbook pages 193–197.

Chemical names and formulas of covalent compounds

1. What is a covalent compound?

____

2. What type of bond is formed in a covalent compound?

____

3. What is used in naming covalent compounds?

____

4. Write the chemical formula for each of the following compounds.

(a) silicon dioxide ____________	(i) dinitrogen pentoxide ____________
(b) chlorine dioxide ____________	(j) dinitrogen monoxide ____________
(c) tellurium dioxide ____________	(k) arsenic tetrabromide ____________
(d) selenium trioxide ____________	(l) arsenic pentachloride ____________
(e) carbon disulphide ____________	(m) disulphide pentoxide ____________
(f) arsenic trichloride ____________	(n) sulphur monochloride ____________
(g) chlorine heptoxide ____________	(o) phosphorus trichloride ____________
(h) selenium difluoride ____________	(p) diphosphorus pentoxide ____________

5. Complete the following crossword puzzle. Given the chemical formula, what is the name fovr the covalent compound?

COVALENT COMPOUNDS

Word List
Arsenic trioxide
Boron monoxide
Carbon disulphide
Chlorine monoxide
Chlorine trifluoride
Diarsenic pentoxide
Dichlorine heptoxide
Dinitrogen trioxide
Diphosphorus trioxide
Disulphur dichloride
Iodine trichloride
Nitrogen dioxide
Nitrogen monoxide
Phosphorus tribromide
Silicon tetrafluoride
Sulphur tetrachloride
Tellurium dibromide
Tellurium trioxide

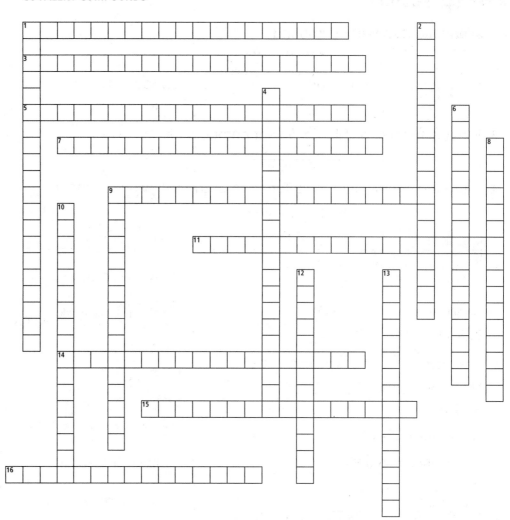

ACROSS
1. S_2Cl_2
3. PBr_3
5. SiF_4
7. Cl_2O_7
9. ClF_3
11. N_2O_3
14. $TeBr_2$
15. ClO
16. AsO_3

DOWN
1. P_2O_3
2. As_2O_5
4. SCl_4
6. ICl_3
8. NO
9. CS_2
10. TeO_3
12. BO
13. NO_2

Use with textbook pages 184–197.

Names and formulas of compounds

Match each Chemical Name on the left with the correct Chemical Formula on the right.	
Chemical Name	**Chemical Formula**
1. _____ tin(II) chlorate **2.** _____ sulphur dichloride **3.** _____ strontium perchlorate	**A.** SCl **B.** S_2Cl **C.** SCl_2 **D.** SnClO **E.** $Sn(ClO_2)_2$ **F.** $Sn(ClO_3)_2$ **G.** $Sn(ClO_4)_2$ **H.** $Sr(ClO_3)_2$ **I.** $Sr(ClO_4)_2$

4. Which of the following is a covalent compound?

    A. SrO

    B. SeO_2

    C. SnO_2

    D. Sc_2O_3

5. In which of the following do covalent bonds hold the atoms together?

    A. silver

    B. calcium carbonate

    C. silicon tetrafluoride

    D. magnesium bromide

6. What is the total number of atoms that make up iodine pentachloride?

    A. 2

    B. 4

    C. 5

    D. 6

7. Which of the following occurs when carbon forms a compound with oxygen?

    A. oxygen and carbon share electrons

    B. both oxygen and carbon lose electrons

    C. oxygen gains electrons, while carbon loses electrons

    D. carbon gains electrons, while oxygen loses electrons

8. In the chemical reaction $CuO + CO_2 \rightarrow CuCO_3$, which of the following are ionic compounds?

I.	CO_2
II.	CuO
III.	$CuCO_3$

    A. I and II only   **C.** II and III only

    B. I and III only   **D.** I, II, and III

9. Which of the following is the formula for the compound formed by ammonium and dichromate?

    A. $NH_4Cr_2O_7^-$

    B. $(NH_4)_2CrO_4$

    C. $NH_4(Cr_2O_7)_2$

    D. $(NH_4)_2Cr_2O_7$

10. In which of the following compounds does manganese have the highest ion charge?

    A. MnO_3

    B. $MnBr_2$

    C. $MnSO_3$

    D. $Mn(OH)_4$

11. In which of the following compounds is the ion charge on copper the same?

I.	Cu_2O
II.	$CuCl_2$
III.	$CuCO_3$

    A. I and II only   **C.** II and III only

    B. I and III only   **D.** I, II, and III

12. In the name arsenic(III) chloride, what does the Roman numeral reveal about arsenic?

    A. it has an ion charge of 3–

    B. it has an ion charge of 3+

    C. it has gained three electrons

    D. it can form three positive ions

Chemical Equations

Textbook pages 202–215

Before You Read

What do you already know about chemical equations? Write your ideas in the lines below.

_____

_____

 Create a Table

Create a table that outlines the steps you need to take when writing and balancing chemical equations

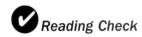 **Reading Check**

List the four states of matter.

_____

_____

_____

_____

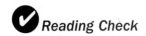 **Reading Check**

What does the law of conservation of mass state?

_____

_____

_____

_____

How are chemical changes and chemical reactions linked?

A chemical change is a change in the arrangements and connections between ions and atoms. Chemical change always involves the conversion of pure substances (elements and compounds) called **reactants** into other pure substances called **products**, which have different properties than the reactants. One or more chemical changes that occur at the same time are called a **chemical reaction**.

How is a chemical reaction represented?

A chemical reaction can be represented using a **chemical equation**. A chemical equation may be written in words (a **word equation**) or in chemical symbols (a **symbolic equation**). In a chemical equation, the reactants are written to the left of an arrow and the products are written to the right. The symbols for **states of matter** may be used to show whether each reactant or product is solid (s), liquid (l), gas (g), or aqueous (aq).

$$2H_2(g) \quad + \quad O_2(g) \quad \rightarrow \quad 2H_2O(g)$$

Chemical reactions obey the law of **conservation of mass**. Atoms are neither destroyed nor produced in a chemical reaction. The total mass of the products is always equal to the total mass of the reactants. ✔

How are chemical equations written and balanced?

Chemical equations are written and balanced through a series of steps, as shown below.

1. Write a word equation: The simplest form of a chemical equation is a word equation. A word equation provides the names of the reactants and products in a chemical reaction. It provides the starting point for writing and balancing chemical equations.

word equation: methane + oxygen → water + carbon dioxide

2. Write a **skeleton equation**: A **skeleton equation** replaces the names of the reactants and products in a word equation with formulas. However, it does not show the correct proportions in which the reactants will actually combine and the products will be produced.
A skeleton equation is not balanced.

skeleton equation: $CH_4 + O_2 → H_2O + CO_2$

3. Write a balanced equation: A **balanced chemical equation** shows the identities of each pure substance involved in the reaction, as well as the number of atoms of each element on both sides of a chemical equation. Chemical equations are balanced using the lowest whole number **coefficients**. These are integers placed in front of the formula or chemical symbol for each product and reactant. The number of atoms after a chemical reaction is the same as it was before a chemical reaction. You can use this information to determine the coefficients that balance the equation.

balanced chemical equation: $CH_4 + 2O_2 → 2H_2O + CO_2$

Name

Date

Section
4.3
Summary

continued

The following strategies can help you translate a word equation into a skeleton equation.

♦ A chemical symbol is used for nearly all elements when they are not in a compound.

♦ Three common compounds containing hydrogen that you should memorize are methane (CH_4), ammonia (NH_3), and water (H_2O).

There are seven common diatomic elements, all of which are non-metals. These are hydrogen, nitrogen, oxygen, fluorine, chlorine, bromine, and iodine. When they occur alone (not in a compound), they are written as H_2, O_2, F_2, Br_2, I_2, N_2, and Cl_2. You can use the word "HOFBrINCl" to remember them. If an element occurs alone and is not diatomic, no subscript is used. For example, in a chemical equation, oxygen is written as O_2 when it occurs alone, while lead is written as Pb.

The following strategies can help you balance a skeleton equation.

♦ Balance compounds first and single elements last.

♦ If you place a coefficient in front of a formula, be sure to balance all the atoms of that formula before moving on.

♦ Add coefficients only in front of formulas. Do not change subscripts.

♦ When oxygen or hydrogen appears in more than one formula on the reactant side or the product side of the chemical equation, balance oxygen and hydrogen last.

♦ You can often treat polyatomic ions, such as SO_4^{2-}, as a unit.

♦ If an equation is balanced by using half a molecule (i.e., ½ O_2), you must double all coefficients so that they are all integers.

♦ When you are finished, perform a final check to be sure that all elements are balanced.

Use with textbook pages 206–211.

Balancing equations

Starting with the skeleton equations, balance the following equations by adding coefficients where appropriate.

1. $H_2 + F_2 \rightarrow HF$ _____

2. $Sn + O_2 \rightarrow SnO$ _____

3. $MgCl_2 \rightarrow Mg + Cl_2$ _____

4. $KNO_3 \rightarrow KNO_2 + O_2$ _____

5. $BN + F_2 \rightarrow BF_3 + N_2$ _____

6. $CuI_2 + Fe \rightarrow FeI_2 + Cu$ _____

7. $Li + H_2O \rightarrow LiOH + H_2$ _____

8. $NH_3 + O_2 \rightarrow N_2 + H_2O$ _____

9. $V_2O_5 + Ca \rightarrow CaO + V$ _____

10. $C_9H_6O_4 + O_2 \rightarrow CO_2 + H_2O$ _____

11. $H_2S + PbCl_2 \rightarrow PbS + HCl$ _____

12. $C_3H_7OH + O_2 \rightarrow CO_2 + H_2O$ _____

13. $Zn + CuSO_4 \rightarrow Cu + ZnSO_4$ _____

14. $C_6H_{12}O_6 + O_2 \rightarrow CO_2 + H_2O$ _____

15. $C_2H_5OH + O_2 \rightarrow CO_2 + H_2O$ _____

16. $Al + H_2SO_4 \rightarrow H_2 + Al_2(SO_4)_3$ _____

17. $FeCl_3 + Ca(OH)_2 \rightarrow Fe(OH)_3 + CaCl_2$ _____

18. $Pb(NO_3)_2 + K_2CrO_4 \rightarrow PbCrO_4 + KNO_3$ _____

19. $Cd(NO_3)_2 + (NH_4)_2S \rightarrow CdS + NH_4NO_3$ _____

20. $Ca(OH)_2 + NH_4Cl \rightarrow NH_3 + CaCl_2 + H_2O$ _____

Name Date

Applying Knowledge
Section 4.3

Use with textbook pages 202–211.

Word equations

Write the skeleton equation for each of the following reactions. Then balance each of the following chemical equations.

1. hydrogen + oxygen → water

2. iron(III) oxide + hydrogen → water + iron

3. sodium + water → sodium hydroxide + hydrogen

4. calcium carbide + oxygen → calcium + carbon dioxide

5. potassium iodide + chlorine → potassium chloride + iodine

6. chromium + tin(IV) chloride → chromium(III) chloride + tin

7. magnesium + copper(II) sulphate → magnesium sulphate + copper

8. zinc sulphate + strontium chloride → zinc chloride + strontium sulphate

9. ammonium chloride + lead(III) nitrate → ammonium nitrate + lead(III) chloride

10. iron(III) nitrate + magnesium sulphide → iron(III) sulphide + magnesium nitrate

11. aluminum chloride + sodium carbonate → aluminum carbonate + sodium chloride

12. sodium phosphate + calcium hydroxide → sodium hydroxide + calcium phosphate

Use with textbook pages 202–203, 206–211.

Chemical reactions and chemical equations

Rewrite the following sentences as chemical word equations. Then write the skeleton equation and balance the equation.

1. Iron combines with oxygen to form rust, which is also known as iron(II) oxide.

Word equation: _____

Balanced equation: _____

2. A solution of hydrogen chloride reacts with sodium carbonate to produce carbon dioxide, sodium chloride, and water.

Word equation: _____

Balanced equation: _____

3. When aluminum metal is exposed to oxygen, a metal oxide called aluminum oxide is formed.

Word equation: _____

Balanced equation: _____

4. Water reacts with powered sodium oxide to produce a solution of sodium hydroxide.

Word equation: _____

Balanced equation: _____

5. Hydrogen gas reacts with nitrogen trifluoride gas to form nitrogen gas and hydrogen fluoride.

Word equation: _____

Balanced equation: _____

6. Chromium(III) sulphate reacts with potassium carbonate to form chromium(III) carbonate and potassium sulphate.

Word equation: _____

Balanced equation: _____

7. Potassium chlorate when heated becomes oxygen gas and potassium chloride.

Word equation: _____

Balanced equation: _____

8. A piece of metallic zinc is placed in a blue solution of copper(II) sulphate. A reddish brown layer of metallic copper forms in a clear solution of zinc sulphate.

Word equation: _____

Balanced equation: _____

Use with textbook pages 202–211.

Chemical equations

Match the Term on the left with the best Descriptor on the right. Each Descriptor may be used only once.	
Term	**Descriptor**
1. _____ product **2.** _____ reactant **3.** _____ coefficient **4.** _____ word equation **5.** _____ skeleton equation **6.** _____ chemical reaction **7.** _____ chemical equation	**A.** a chemical that reacts in a chemical reaction **B.** a chemical that forms in a chemical reaction **C.** a chemical change in which new substances are formed **D.** a chemical equation that is written using chemical names **E.** an integer placed in front of a formula in a chemical equation **F.** a chemical equation that is written using chemical formulas **G.** a set of chemical formulas that identify the reactants and products in a chemical reaction

8. Which of following describes the law of conservation of mass?

I.	The mass is conserved in a chemical reaction.
II.	The total mass of the products is equal to the total mass of the reactants in a chemical reaction.
III.	The total number of each kind of atom at the start of the reaction is equal to the total number of each kind of atom after the reaction.

A. I and II only

B. I and III only

C. II and III only

D. I, II, and III

9. How many oxygen atoms are there in the compound lead(IV) bisulphate, $Pb(HSO_4)_4$?

A. 2   **C.** 8

B. 4   **D.** 16

10. Which of the following are diatomic elements?

I.	iodine
II.	nitrogen
III.	hydrogen

A. I and II only   **C.** II and III only

B. I and III only   **D.** I, II, and III

Use the following unbalanced equation to answer question 11.

$$PCl_5 + H_2O \rightarrow HCl + H_3PO_4$$

11. Which of the following sets of coefficients will balance the equation?

A. 1, 4, 5, 1   **C.** 1, 3, 5, 2

B. 1, 5, 4, 1   **D.** 1, 4, 2, 1

12. A solution of sodium sulphide is mixed with a solution of copper(II) nitrate. A precipitate of copper sulphide is formed in a solution of sodium nitrate. What are the reactants in this chemical reaction?

A. Na_2S and CuS

B. CuS and $NaNO_3$

C. Na_2S and $Cu(NO_3)_2$

D. Na_2SO_4 and Cu_2NO_3

13. A piece of aluminum metal is placed in a solution of sulphuric acid, H_2SO_4. A compound, aluminum sulphate, forms and bubbles are seen going to the surface. What type of gas formed during this reaction?

A. oxygen   **C.** carbon dioxide

B. hydrogen   **D.** carbon monoxide

Acids and Bases

Textbook pages 220–233

Before You Read

Many acids and bases can be found in your home. Describe one acid and one base that you are familiar with. Record your answer in the lines below.

_____

_____

What are acids and bases?

Many common pure substances can be classified according to whether they are acids or bases. Acids produce **hydrogen ions** (H⁺) and bases produce **hydroxide ions** (OH⁻) when dissolved in solution. The **concentration** of hydrogen ions refers to the number of hydrogen ions in a specific volume of solution. Solutions with a high concentration of hydrogen ions are highly acidic. Similarly, solutions with a high concentration of hydroxide ions are highly basic. When an acidic solution is mixed with a basic solution, the solutions can **neutralize** each other, which means that the acidic and basic properties are in balance.

What is pH?

Testing the pH of a solution is a way of measuring its concentration of hydrogen ions, H⁺(aq). The **pH scale** is a number scale that indicates how acidic or basic a solution is. **Acids** have a pH below 7 and **bases** have a pH above 7. Neutral solutions have a pH of 7. On the pH scale, one unit of change represents a 10-fold change in the degree of acidity or basicity. For example, a two unit drop in pH is a 10^2 or 100 times increase in acidity.

What are pH indicators?

pH indicators are chemicals that change colour depending on the pH of a solution.

- ◆ **Litmus paper** can determine whether a solution is acidic or basic. Blue litmus paper turns red in an acidic solution (below pH 7). Red litmus paper changes to blue in a basic solution (above pH 7),

Create a Quiz

After you have read this section, create a five-question quiz on acids. Answer your questions until you get them all correct.

✔ **Reading Check**

What is the pH of a neutral solution?

_____

♦ A universal indicator contains a number of indicators that turn different colours depending on the pH of the solution.

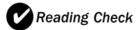
Reading Check

Provide the name of a common pH indicator.

_____

♦ **Phenolphthalein, bromothymol blue, indigo carmine, methyl orange**, and **methyl red** are other common pH indicators. Each determines pH within a different range.

A digital pH meter or pH computer probe that measures the electrical properties of a solution can also be used to determine pH.

How are acids and bases named?

Generally, the chemical formula for an acid starts with H (hydrogen) on the left hand side of the formula. Acids can be named in several ways. Many compounds take on acidic properties only when mixed with water. If no state of matter is given, the name may be begin with hydrogen, as in hydrogen chloride (HCl). However, if the acid is shown as being aqueous (dissolved in water), a different name may be used—one that ends in **-ic acid**, as in hydrochloric acid. Other acids that do not contain oxygen, such as hydrofluoric acid, HF(aq); nitric acid, HNO_3 (aq); and sulphuric acid, H_2SO_4(aq), also follow this naming system.

Another naming system is followed when oxygen is present in the formula. Names that begin with hydrogen and end with the suffix **-ate** (i.e., hydrogen carbonate, H_2CO_3) can be changed by dropping "hydrogen" from the name and changing the suffix to **-ic acid** (i.e., carbonic acid, H_2CO_3(aq)). When the name begins with hydrogen and ends with the suffix **-ite** (i.e., hydrogen sulphite, H_2SO_3), then the name can be changed to end with the suffix **-ous acid** (i.e., sulphurous acid, H_2SO_3(aq)).

Bases generally have OH on the right hand side of their chemical formulas. Common names of bases include sodium hydroxide (NaOH) and magnesium hydroxide (Mg(OH)$_2$).

What are the properties of acids and bases?

Some of the properties of acids and bases are compared in the table below.

Property	Acid	Base
Taste CAUTION: Never taste chemicals in the laboratory.	• Acids taste sour. Lemons, limes, and vinegar are common examples.	• Bases taste bitter. The quinine in tonic water is one example.
Touch CAUTION: Never touch chemicals in the laboratory with your bare skin.	• Many acids will burn your skin. Sulfuric acid (battery acid) is one example.	• Bases feel slippery. • Many bases will burn your skin. Sodium hydroxide (lye) is one example.
Indicator tests	• Acids turn blue litmus paper red. • Phenolphthalein is colourless in an acidic solution.	• Bases turn red litmus blue. • Phenolphthalein is colourless in slightly basic solutions and pink in moderate to strongly basic solutions.
Reaction with some metals, such as magnesium or zinc	• Acids corrode metals.	• No reaction
Electrical conductivity	• Conductive	• Conductive
pH	• Less than 7	• More than 7
Production of ions	• Acids form hydrogen (H^+) ions when dissolved in solution.	• Bases form hydroxide (OH^-) ions when dissolved in solution.

What are some common acids and bases?

Formula	Name	Examples of uses
CH_3COOH	ethanoic acid or acetic acid	in vinegar
H_2SO_4	sulphuric acid	automobile batteries
$NaOH$	sodium hydroxide	drain and overn cleaners
$Mg(OH)_2$	magnesium hydroxide	antacids
HCl	hydrochloric acid	digestion in stomach

Use with textbook pages 220–224.

pH scale and pH indicators

1. Define the following terms:

(a) pH indicator _____

(b) pH scale _____

Figure 1: pH values of common substances

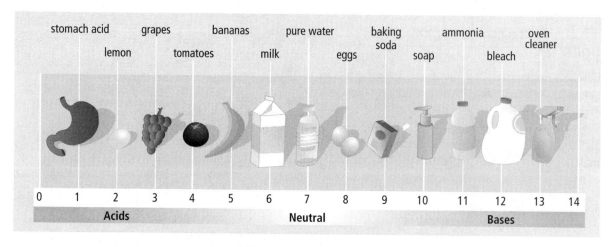

Figure 2: Common acid-base indicators and their pH colour change

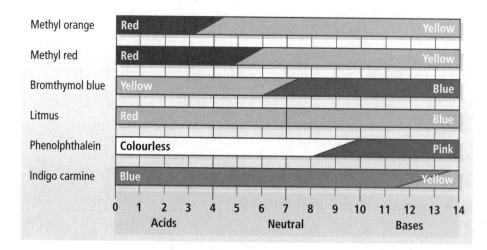

2. Complete the following tables by using the two figures shown on the previous page. Identify whether the substance is an acid or a base and indicate what colour the pH indicator will turn.

(a)

Substance	pH Value	Acid or Base	Methyl Orange	Bromothymol Blue	Litmus
lemon					
ammonia					
milk					

(b)

Substance	pH Value	Acid or Base	Methyl Red	Phenolphthalein	Indigo Carmine
tomato					
oven cleaner					
egg					

3. Complete the following table. Identify whether the substance is an acid or a base and indicate what colour the pH indicator will turn.

Substance	pH Value	Acid or Base	pH Indicator	Colour of pH Indicator
black coffee	5		litmus	
milk of magnesia	10		phenolphthalein	
battery acid	0		bromothymol blue	
sea water	8		indigo carmine	
orange juice	3		methyl orange	
liquid drain cleaner	14		methyl red	

Use with textbook pages 225–226.

Names of acids

1. An acid will have the suffix "–ic acid" at the end of its name when the negative ion has a suffix _____. For example, "hydrogen carbon**ate** (H_2CO_3)" is called "carbon**ic acid**".

2. An acid will have the suffix "–ous acid" at the end of its name when the negative ion has a suffix _____. For example, "hydrogen sulph**ite** (H_2SO_3)" is called "sulphur**ous acid**."

3. What is the name of each of the following acids?

(a) H_2CO_3 _____

(b) CH_3COOH _____

(c) H_3PO_4 _____

(d) $HClO_2$ _____

(e) H_2SO_3 _____

(f) HNO_3 _____

(g) HF _____

(h) HCl _____

4. What is the chemical formula for each of the following acids?

(a) hydriodic acid _____

(b) sulphuric acid _____

(c) perchloric acid _____

(d) nitrous acid _____

(e) chloric acid _____

(f) hydrobromic acid _____

(g) phosphorous acid _____

(h) hypochlorous acid _____

Use with textbook pages 220–229.

Acids versus bases

1. Compare and contrast acids and bases by completing the following table.

	Acids	Bases
definition		
pH		
what to look for in chemical formula		
production of ions		
electrical conductivity		
taste		
touch		
examples		

2. Classify each of the following as an acid or a base.

(a) H_3PO_4 _____

(b) NH_4OH _____

(c) $Mg(OH)_2$ _____

(d) has a pH of 4 _____

(e) has a pH of 9 _____

(f) sulphurous acid _____

(g) hydrogen bromide _____

(h) potassium hydroxide _____

(i) causes methyl orange to turn red _____

(j) causes phenolphthalein to turn pink _____

(k) causes indigo carmine to turn yellow _____

(l) causes bromothymol blue to turn yellow _____

Use with textbook pages 220–229.

Acids and bases

Match the Term on the left with the best Descriptor on the right. Each Descriptor may be used only once.	
Term	**Descriptor**
1. _____acid 2. _____base 3. _____neutral 4. _____pH scale 5. _____corrosive 6. _____pH indicator 7. _____concentration of hydrogen	**A.** a solution with a pH of 7 **B.** can burn skin or eyes on contact **C.** number of hydrogen ions in a specific volume of solution **D.** a chemical compound that produces a solution with a pH less than 7 **E.** a number scale for measuring how acidic or basic a solution is **F.** a chemical compound that produces a solution with a pH greater than 7 **G.** a chemical that changes colour depending on the pH of the solution it is placed in

8. Which of the following describes acids?

I.	has a pH of less than 7
II.	can conduct electricity
III.	produce hydroxide ions when dissolved in solution

A. I and II only

B. I and III only

C. II and III only

D. I, II, and III

9. What happens to the number of H^+ after H_2SO_4 is added to water?

A. it increases

B. it decreases

C. it stays the same

10. Which of the following is a base?

A. KCl **C.** LiOH

B. HBr **D.** HNO_3

11. What is the name for $HClO_3$?

A. chloric acid

B. chlorous acid

C. perchloric acid

D. hypochlorous acid

12. What is the chemical formula for sulphurous acid?

A. HS **C.** H_2SO_3

B. HSO_4^- **D.** H_2SO_4

13. What is the pH of a substance that causes methyl orange to turn yellow and methyl red to turn red?

A. 3 **C.** 6.5

B. 4.5 **D.** 8

14. Which of the following would occur if eggs were tested with various pH indicators?

I.	indigo carmine turns blue
II.	phenolphthalein turns pink
III.	bromothymol blue turns blue

A. I and II only

B. I and III only

C. II and III only

D. I, II, and III

Salts

Textbook pages 234–243

Before You Read

How many different uses for salts can you name? Write your answers on the lines below.

_____

_____

What are salts?

In chemistry, **salts** are a class of ionic compounds that can be formed during the reaction of an acid and a base. A salt is made up of a positive ion from a base and a negative ion from an acid. An acid and a base react to form a salt and water in a chemical reaction called a **neutralization (acid-base)** reaction. For example:

$$HCl + NaOH \rightarrow NaCl + H_2O$$
acid +   base → salt   + water ✔

Which other compounds react with acids to produce salts?

Acids can also react with metals and carbonates to produce salts

1. Metals: When metals react with acids to produce a salt, they usually release hydrogen gas, as shown below.

   $$2HCl(aq) + Mg(s) \rightarrow MgCl_2(aq) + H_2(g)$$

   The most reactive metals are the **alkali metals** and alkaline **earth metals**, which appear on the extreme left of the periodic table. Within these groups, the elements at the bottom of the columns react the most vigorously.

2. Carbonates: Carbonates can also react with acids to produce salts. Much of the carbon dioxide on the surface of Earth is trapped in rocks, such as limestone, dolomite, and calcite, which contain carbonate ions. When carbonate rocks react with acids, such as those in acid precipitation, the carbonates help to neutralize the acid. Sulphuric acid is one component of acid precipitation.

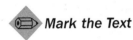
Mark the Text

Identify Concepts

Highlight each question head in this section. Then use a different colour to highlight the answers to the questions.

✔ Reading Check

What is a salt?

_____

_____

_____

_____

The chemical reaction between this acid and carbonate releases carbon dioxide gas, as shown below.

$$H_2SO_4 + CaCO_3 \rightarrow CaSO_4 + H_2O + CO_2$$

What are oxides?

An **oxide** is a chemical compound that includes at least one oxygen atom or ion along with one or more other elements. Both metals and non-metals can form oxides. ✔

✔ Reading Check

Which type of oxide combines with water to form a base?

_____

1. Metal oxides: A **metal oxide** is a chemical compound that contains a metal chemically combined with oxygen. A metal oxide, such as sodium oxide, combines with water to form a base (see below).

   $$Na_2O(s) + H_2O(l) \rightarrow 2NaOH(aq)(a\ base)$$
   $$\qquad\qquad\qquad\qquad sodium\ hydroxide$$

   The base can then react chemically with an acid to form a salt.

2. Non-metal oxides: A **non-metal oxide** is a chemical compound that contains a non-metal chemically combined with oxygen. A non-metal oxide, such as carbon dioxide, combines with water to form an acid.

   $$CO_2(g) + H_2O(l) \rightarrow H_2CO_3(aq)$$
   $$\qquad\qquad\qquad carbonic\ acid$$

   This acid can react chemically with a base to form a salt.

Use with textbook pages 234–239.

Recognizing acids, bases, and salts

1. State whether each of the following is an acid, a base, or a salt.

(a) HI _____

(b) HBr _____

(c) KOH _____

(d) HNO_3 _____

(e) NaOH _____

(f) H_2SO_4 _____

(g) H_2CO_3 _____

(h) H_3PO_4 _____

(i) Na_3PO_4 _____

(j) $Sr(OH)_2$ _____

(k) $Ca(OH)_2$ _____

(l) $Al_2(SO_4)_3$ _____

(m) CH_3COOH _____

(n) $Mg(CH_3COO)_2$ _____

(o) calcium nitrate _____

(p) sodium chloride _____

(q) hydrocyanic acid _____

(r) hydrogen fluoride _____

(s) barium hydroxide _____

(t) hypochlorous acid _____

(u) aluminum hydroxide _____

(v) magnesium carbonate _____

2. What acid is present in vinegar? _____

3. What is the chemical name for table salt? _____

4. What acid is used in automobile batteries? _____

5. What base is found in drain and oven cleaners? _____

6. What base is the active ingredient in some antacids? _____

7. What acid is produced in the stomach to help digest food? _____

Name ____________________ Date ____________________

Analyzing Information
Section 5.2

Use with textbook pages 234–239.

Acid-base neutralization reactions

acid + base → water + salt

1. Complete and balance the following neutralization reactions.

   (a) H_2SO_4 + NaOH → ____________________

   (b) HNO_3 + KOH → ____________________

   (c) HCl + $Ca(OH)_2$ → ____________________

   (d) H_3PO_4 + $Ba(OH)_2$ → ____________________

   (e) CH_3COOH + NaOH → ____________________

   (f) HNO_3 + $Sr(OH)_2$ → ____________________

   (g) HF + $Fe(OH)_3$ → ____________________

   (h) HBr + $Sn(OH)_4$ → ____________________

2. Complete and balance the following acid-base neutralization reactions. Include both the word equation and the formula.

   (a) sulphuric acid + potassium hydroxide → ____________________

   (b) acetic acid + barium hydroxide → ____________________

   (c) phosphoric acid + aluminum hydroxide → ____________________

   (d) nitric acid + lithium hydroxide → ____________________

   (e) sulphuric acid + calcium hydroxide → ____________________

   (f) hydrochloric acid + magnesium hydroxide → ____________________

Use with textbook pages 237.

Metal oxides and non-metal oxides

1. What element reacts with a metal or a non-metal to form an oxide?

_____

2. What is a chemical compound that contains a metal chemically combined with oxygen? _____

3. What is a chemical compound that contains a non-metal chemically combined with oxygen? _____

4. What happens to a solution when a metal oxide dissolves in water?

_____

5. What happens to a solution when a non-metal oxide dissolves in water?

_____

6. What is formed when a metal oxide reacts with water? _____

7. What is formed when a non-metal oxide reacts with water?

_____

8. Classify each of the following as a metal oxide or a non-metal oxide.

(a) Na_2O _____ (e) SO_2 _____

(b) B_2O_3 _____ (f) BeO _____

(c) NO_2 _____ (g) ClO _____

(d) CaO _____ (h) Li_2O _____

9. Indicate whether an acid or a base will be produced.

(a) $MgO + H_2O \rightarrow$ _____

(b) $SO_3 + H_2O \rightarrow$ _____

(c) $CaO + H_2O \rightarrow$ _____

(d) $CO_2 + H_2O \rightarrow$ _____

Use with textbook pages 234–239.

Salts

Match the Term on the left with the best Chemical Formula on the right. Each Chemical Formula may be used only once.	
Term	**Chemical Formula**
1. _____ water	**A.** H_2O
2. _____ a salt	**B.** NO_2
3. _____ a base	**C.** $MgCl_2$
4. _____ an acid	**D.** Na_2O
5. _____ a metal oxide	**E.** H_2CO_3
6. _____ a non-metal oxide	**F.** NH_4OH

7. Which of the following metals is most reactive?

   A. copper

   B. sodium

   C. francium

   D. magnesium

8. When non-metal oxides dissolve in water, the solution becomes

   A. basic

   B. acidic

   C. neutral

9. Carbon dioxide forms which of the following in water?

   A. CO

   B. CO_3^{2-}

   C. HCO_3^-

   D. H_2CO_3

10. What coefficient is needed for sodium hydroxide in order to balance the following equation?

   $$H_2SO_4 + NaOH \rightarrow Na_2SO_4 + H_2O$$

   A. 1   **C.** 3

   B. 2   **D.** 4

11. Hydrochloric acid can be used to neutralize potassium hydroxide. What is the formula for the salt produced by this neutralization?

   A. H_2O

   B. KCl

   C. $KClO_2$

   D. $KClO_3$

12. Which reactants form the salt $FePO_4$ in a neutralization reaction?

   A. PO_4 and Fe_2O_3

   B. H_3P and $Fe(OH)_3$

   C. H_2O and $Fe(OH)_3$

   D. H_3PO_4 and $Fe(OH)_3$

Use the following acid-base neutralization reaction to answer question 13.

$$H_2CO_3 + Ba(OH)_2 \rightarrow BaCO_3 + 2 H_2O$$

13. Which of the following statements is true?

I.	H_2CO_3 is an acid.
II.	$BaCO_3$ is a base.
III.	The products of this reaction are a salt and water.

   A. I and II only

   B. I and III only

   C. II and III only

   D. I, II, and III

Organic Compounds

Textbook pages 244–251

Before You Read

What do you think of when you hear the term "organic"? Outline your thoughts in the lines below.

_____

_____

 Make Flash Cards

Create flash cards to help you remember common organic compounds. Write the name of the compound on the front of the card and the information you want to recall on the back.

What are organic compounds?

Organic compounds are any compounds that contain carbon (with a few exceptions). All other compounds are referred to as **inorganic compounds**. In almost all organic compounds, carbon atoms are bonded to hydrogen atoms or other elements that are near carbon in the periodic table, especially nitrogen, oxygen, sulphur, phosphorus, and the halogens. Other elements, including metals and non-metals, may also be present.

The carbon in organic compounds forms four bonds, which enables it to form complex, branched-chain structures, ring structures, and even cage-like structures. Several different methods can be used to model these structures. These include the structural formula, the ball-and-stick model, and the space-filling model shown below.

CH_4

molecular formula

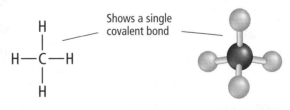

Shows a single covalent bond

structural formula   ball-and-stick model

space-filling model

To recognize a compound as organic, look for an indication of the presence of carbon in its name, chemical formula, or diagram. However, there are a few exceptions to this rule. Certain compounds that contain carbon are classified as inorganic carbon compounds. These include any compounds that contain carbonates, (i.e., $CaCO_3$); carbides, (i.e., SiC); and oxides (i.e., CO_2, CO). ✔

What are some common organic compounds?

Two common organic compounds are hydrocarbons and alcohols.

1. Hydrocarbons: A **hydrocarbon** is an organic compound that contains only the elements carbon and hydrogen. The simplest of all organic compounds is the hydrocarbon molecule called methane (CH_4) which consists of a carbon atom bonded to four hydrogen atoms. Other hydrocarbons are formed by linking two or more carbons together to make a chain. The first five hydrocarbons are given in the table below.

✔ **Reading Check**

How does an organic compound differ from an inorganic compound?

_____

_____

_____

Name	Molecular Formula	Structural Formula	Shortened Structural Formula	Space-Filling Model	Common Uses
methane	CH_4	H │ H─C─H │ H	CH_4		• Natural gas heaters
ethane	C_2H_6	H H │ │ H─C─C─H │ │ H H	CH_3CH_3		• Manufacturing plastic
propane	C_3H_8	H H H │ │ │ H─C─C─C─H │ │ │ H H H	$CH_3CH_2CH_3$		• Camp fuel
butane	C_4H_{10}	H H H H │ │ │ │ H─C─C─C─C─H │ │ │ │ H H H H	$CH_3CH_2CH_2CH_3$		• Hand-held lighters
pentane	C_5H_{12}	H H H H H │ │ │ │ │ H─C─C─C─C─C─H │ │ │ │ │ H H H H H	$CH_3CH_2CH_2CH_2CH_3$		• Component of gasoline

✔ Reading Check

Provide the molecular
formula for ethanol.

_____

2. Alcohols: An **alcohol** is one kind of organic compound
that contains C, H, and O in a specific structure. The
table below shows some common alcohols. ✔

Name	Molecular Formula	Structural Formula	Shortened Structural Formula	Space-Filling Model	Common Use
methanol	CH_4O	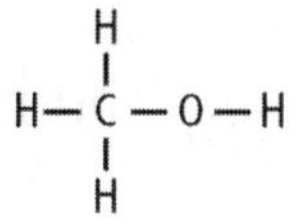	CH_3OH		• Solvent
ethanol	C_2H_6O	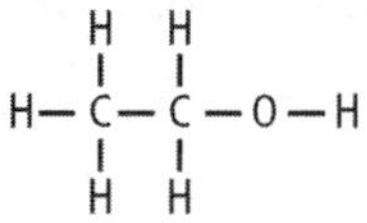	CH_3CH_2OH		• Fuel
isopropyl alcohol	C_3H_8O	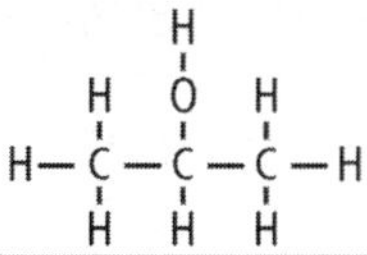	$(CH_3)CH_2OH$		• Sterilizer • Cleaner

Use with textbook pages 244–248.

Organic chemistry

Vocabulary		
alcohol	ethanol	organic chemistry
butane	hydrocarbons	organic compounds
carbon	inorganic compounds	oxygen
ethane	methane	propane
		solvent

Use the terms in the vocabulary box to fill in the blanks. You may use each term only once.

1. Almost all compounds that contain carbon, with the exception of carbon dioxide, carbon monoxide, and ionic carbonates, are _____. The study of carbon-containing compounds is known as _____.

2. _____ are compounds that do not contain carbon.

3. _____ is an element with an atomic number of 6. It has four electrons in its valence shell and can form four covalent bonds.

4. Compounds that contain only hydrogen atoms and carbon atoms are called _____.

5. _____, CH_4, is the simplest hydrocarbon, with four hydrogens covalently bonded to one carbon. It is a gas at room temperature.

6. _____, C_2H_6, is a gas at room temperature and is used in manufacturing plastic.

7. _____, C_3H_8, is a gas that is easily turned into a liquid under pressure. That is why it is often used as fuel for camp stoves and gas-fired barbeques.

8. _____, C_4H_{10}, is a gas that is used in hand-held lighters.

9. An _____, such as isopropyl alcohol, is a compound that contains carbon, hydrogen, and _____.

10. Methanol is an example of a _____, which is a liquid that can dissolve other substances.

11. _____, an alcohol with the formula of C_2H_6O or C_2H_5OH, can be seen to be related to the hydrocarbon ethane, C_2H_6, if one H is removed and replaced with OH.

Name ______________________ Date ______________________

Comprehension
Section 5.3

Use with textbook pages 244–248.

Recognizing organic and inorganic compounds

Classify each of the following compounds as organic or inorganic by examining their formulas.

1. CO ______
2. CH_4 ______
3. HCl ______
4. NH_3 ______
5. CO_2 ______
6. CrS ______
7. C_2H_4 ______
8. C_4H_{10} ______
9. C_6H_{14} ______
10. C_8H_{18} ______
11. Cu_2O ______
12. Cr_2O_3 ______
13. $CHCl_3$ ______
14. $CaCO_3$ ______
15. C_2H_6O ______
16. CH_3OH ______
17. $NaHCO_3$ ______
18. $C_6H_{12}O_6$ ______
19. Na_2CO_3 ______
20. $K_2Cr_2O_7$ ______
21. $Ca(OH)_2$ ______
22. $Co(NO_3)_2$ ______
23. $C_{19}H_{28}O_2$ ______
24. NH_4OH ______
25. CH_3OCH_3 ______
26. $C_{18}H_{21}NO_3$ ______
27. CH_3COOH ______
28. CH_3NHCH_3 ______
29. CH_3CH_2OH ______
30. $CH_3CH_2OCH_2CH_3$ ______

Use with textbook pages 244–248.

Organic compounds versus inorganic compounds

Classify each of the following compounds as organic or inorganic by examining the structural formula, ball-and-stick model, or space-filling model.

	Structural formula, ball-and-stick model, or space-filling model	Type of compound (Organic or Inorganic)
1.		
2.		
3.		
4.		
5.		
6.		
7.		
8.		

Use with textbook pages 244–248.

Organic compounds

Using the compound ethane, match the Descriptor on the left with the best Formula / Model that represents ethane on the right. Each Formula / Model may be used only once.

Descriptor	Formula / Model
1. _____ structural formula	**A.** C_2H_6
2. _____ molecular formula	**B.** 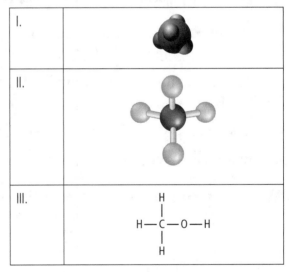
3. _____ space-filling model	**C.**
4. _____ ball-and-stick model	**D.**

5. What element must always be present in an organic compound?

   A. carbon

   B. oxygen

   C. chlorine

   D. hydrogen

6. Which formula represents a hydrocarbon?

   A. $HClO_3$

   B. CH_3COOH

   C. $CH_3CH_2CH_2COOH$

   D. $CH_3CH_2CH_2CH_2CH_3$

7. Which of the following represents an alcohol?

   A.

   B.

   C.

   D.

8. Which of the following represents methane, CH_4?

I.	
II.	
III.	

   A. I and II only

   B. I and III only

   C. II and III only

   D. I, II, and III

Types of Chemical Reactions

Textbook pages 256–271

Before You Read

Many chemical reactions occur in daily life. In the lines below, describe one chemical reaction you have observed.

_____

_____

How are chemical reactions classified?

Chemical reactions can be classified as one of six main types: synthesis, decomposition, single replacement, double replacement, neutralization (acid-base), or combustion. You can identify each type of reaction by examining the reactants. This makes it possible to classify a reaction and then predict the identity of the products.

What is a synthesis (combination) reaction?

In a **synthesis** (combination) reaction, two or more reactants (A and B) combine to produce a single product (AB).

element + element → compound
A + B → AB

(The letters A and B represent elements.)

hydrogen + oxygen → water

What is a decomposition reaction?

In a **decomposition** reaction a compound is broken down into smaller compounds or separate elements. A decomposition reaction is the reverse of a synthesis reaction.

compound → element + element
AB → A + B

calcium chlorate → calcium chloride + oxygen

 Make Flash Cards

Create flash cards to help you learn the different reactions. Write the name of the reaction on the front of the card and an example on the back.

Reading Check

How many products are there in a synthesis reaction?

_____

What is a single replacement reaction?

In a **single replacement** reaction, a reactive element (a metal or a non-metal) and a compound react to produce another element and another compound. In other words, one of the elements in the compound is replaced by another element. The element that is replaced could be a metal or a non-metal.

element + compound → element + compound
A + BC → B + AC where A is a metal OR
A + BC → C + BA where A is a non-metal

aluminum + lead(II) nitrate → aluminum nitrate + lead

What is a double replacement reaction?

A **double replacement** reaction usually involves two ionic solutions that react to produce two other ionic compounds. One of the compounds forms a **precipitate**, which is an insoluble solid that forms from a solution. The precipitate floats in the solution, then settles and sinks to the bottom. The other compound may also form a precipitate, or it may remain dissolved in solution.

ionic solution + ionic solution → ionic solution + ionic solid
AB(aq) + CD(aq) → AD(aq) + CB(s)

iron(II) chloride + lithium phosphate
→ iron(II) phosphate + lithium chloride

What is a neutralization (acid-base) reaction?

When an acid and a base are combined, they will neutralize each other. In a neutralization (acid-base) reaction, an acid and a base react to form a salt and water.

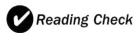 **Reading Check**

What is another name for a neutralization reaction?

_____

_____

acid + base → salt + water
$HX + MOH → MX + H_2O$

(X represents a negative ion. M represents a positive ion.)

sulphuric acid + sodium hydroxide
→ sodium sulphate + water ✔

Name

Date

Section
6.1
Summary

continued

What is a combustion reaction?

Combustion is the rapid reaction of a compound or element with oxygen to form an oxide and produce heat. For example, organic compounds, such as methane, combust with oxygen to form carbon dioxide (the oxide of carbon) and water (the oxide of hydrogen).

hydrocarbon + oxygen ➔ carbon dioxide + water

$$C_XH_Y + O_2 \rightarrow CO_2 + H_2O$$

(The subscripts X and Y represent integers.)

$$C_2H_6O_3 + O_2 \rightarrow CO_2 + H_2O$$

The summary chart below compares the six types of chemical reactions.

Reaction Type	Reactants and Products	Notes on the Reactants
Synthesis (combination)	$A + B \rightarrow AB$	• Two elements combine
Decomposition	$AB \rightarrow A + B$	• One reactant only

Single replacement
 	
  If A is a metal
 	
  If A is a non-metal
	
	$A + BC \rightarrow B + AC$
	
	$A + BC \rightarrow C + BA$
		
		• One element and one compound

Double replacement	$AB + CD \rightarrow AD + CB$	• Two compounds react.
Neutralization (acid-base)	$HX + MOH \rightarrow MX + H_2O$	• Acid plus base
Combustion	$C_XH_Y + O_2 \rightarrow CO_2 + H_2O$	• Organic compound with oxygen

Name ____________________ Date ____________________

Comprehension
Section 6.1

Use with textbook pages 256–267.

Classifying chemical reactions

Classify each of the following reactions as synthesis (S), decomposition (D), single replacement (SR), double replacement (DR), neutralization (N), or combustion (C). Place the correct letter representing the reaction type in the space provided. Then **balance** the chemical equation by placing the correct coefficients in the equation.

_______ 1. ___ N_2 + ___ F_2 → ___ NF_3

_______ 2. ___ $KClO_3$ → ___ KCl + ___ O_2

_______ 3. ___ $C_{12}H_{22}O_{11}$ + ___ O_2 → ___ CO_2 + ___ H_2O

_______ 4. ___ $CuSO_4$ + ___ Fe → ___ $Fe_2(SO_4)_3$ + ___ Cu

_______ 5. ___ MgF_2 + ___ Li_2CO_3 → ___ $MgCO_3$ + ___ LiF

_______ 6. ___ H_3PO_4 + ___ NH_4OH → ___ H_2O + ___ $(NH_4)_3PO_4$

_______ 7. ___ NaF + ___ Br_2 → ___ $NaBr$ + ___ F_2

_______ 8. ___ CH_3OH + ___ O_2 → ___ CO_2 + ___ H_2O

_______ 9. ___ $ZnCl_2$ → ___ Zn + ___ Cl_2

_______ 10. ___ $RbNO_3$ + ___ BeF_2 → ___ $Be(NO_3)_2$ + ___ RbF

_______ 11. ___ S_8 + ___ H_2 → ___ H_2S

_______ 12. ___ $LiCl$ + ___ Br_2 → ___ $LiBr$ + ___ Cl_2

_______ 13. ___ H_2SO_4 + ___ KOH → ___ H_2O + ___ K_2SO_4

_______ 14. ___ $C_{10}H_8$ + ___ O_2 → ___ CO_2 + ___ H_2O

_______ 15. ___ HI → ___ H_2 + ___ I_2

_______ 16. ___ HCl + ___ Al → ___ H_2 + ___ $AlCl_3$

_____ 17. ___ P + ___ Cl_2 → ___ PCl_3

_____ 18. ___ C_6H_6 + ___ O_2 → ___ CO_2 + ___ H_2O

_____ 19. ___ K_2SO_4 + ___ $BaCl_2$ → ___ $BaSO_4$ + ___ KCl

_____ 20. ___ H_3PO_4 + ___ $Ca(OH)_2$ → ___ H_2O + ___ $Ca_3(PO_4)_2$

_____ 21. ___ NF_3 → ___ N_2 + ___ F_2

_____ 22. ___ Al + ___ N_2 → ___ AlN

_____ 23. ___ HF + ___ $Fe(OH)_3$ → ___ H_2O + ___ FeF_3

_____ 24. ___ GaF_3 + ___ Cs → ___ CsF + ___ Ga

_____ 25. ___ $Ca(NO_3)_2$ + ___ Na_3PO_4 → ___ $Ca_3(PO_4)_2$ + ___ $NaNO_3$

_____ 26. ___ HCl + ___ $Al(OH)_3$ → ___ $AlCl_3$ + ___ H_2O

_____ 27. ___ C_5H_{12} + ___ O_2 → ___ CO_2 + ___ H_2O

_____ 28. ___ H_2O_2 → ___ H_2O + ___ O_2

_____ 29. ___ NH_4HCO_3 + ___ NaCl → ___ $NaHCO_3$ + ___ NH_4Cl

_____ 30. ___ Na + ___ O_2 → ___ Na_2O

Name ____________________ Date ____________________

Applying Knowledge
Section 6.1

Use with textbook pages 256–267.

Types of chemical reactions—Word equations

Classify each of the following chemical reactions as synthesis (S), decomposition (D), single replacement (SR), double replacement (DR), or neutralization (N). Then **write a balanced equation** for each word equation.

________ 1. magnesium + sulphur → magnesium sulphide

____

________ 2. potassium hydroxide + sulphuric acid → water + potassium sulphate

____

________ 3. chlorine + potassium iodide → potassium chloride + iodide

____

________ 4. aluminum chloride + sodium hydroxide → aluminum hydroxide + sodium chloride

____

________ 5. lead(II) oxide → lead + oxygen

____

________ 6. magnesium + silver nitrate → silver + magnesium nitrate

____

________ 7. cadmium(II) nitrate + ammonium sulphide → cadmium(II) sulphide + ammonium nitrate

____

________ 8. tin(IV) hydroxide + hydrogen bromide → water + tin(IV) bromide

____

________ 9. sodium + oxygen → sodium oxide

____

________ 10. sodium nitride → sodium + nitrogen

____

_____ 11. calcium hydroxide + phosphoric acid → water + calcium phosphate

_____ 12. barium chloride + sodium carbonate → barium carbonate + sodium chloride

_____ 13. zinc + nickel(II) nitrate → zinc nitrate + nickel

_____ 14. antimony + iodine → antimony(III) iodide

_____ 15. carbon dioxide → carbon + oxygen

_____ 16. iron(III) sulphate + lead → lead(II) sulphate + iron

_____ 17. barium nitrate + ammonium carbonate → ammonium nitrate + barium carbonate

_____ 18. zinc hydroxide + hydrochloric acid → water + zinc chloride

_____ 19. ammonium carbonate + magnesium chloride → ammonium chloride + magnesium carbonate

_____ 20. rubidium hydroxide + sulphuric acid → water + rubidium sulphate

Use with textbook pages 256–267.

Predicting the products

1. For each of the following:

    I. predict the products

    II. classify the reaction as synthesis (S), decomposition (D), single replacement (SR), double replacement (DR), neutralization (N), or combustion (C)

    III. write a balanced equation

    _____ (a) H_2O $\rightarrow$

    _____ (b) H_2 + Cl_2 $\rightarrow$

    _____ (c) NaI + F_2 $\rightarrow$

    _____ (d) $AgNO_3$ + Na_3PO_4 $\rightarrow$

    _____ (e) $Ba(OH)_2$ + H_2SO_4 $\rightarrow$

    _____ (f) P_4 + Cl_2 $\rightarrow$

    _____ (g) CH_3OH + O_2 $\rightarrow$

    _____ (h) $Sr(OH)_2$ + H_3PO_4 $\rightarrow$

    _____ (i) FeI_2 $\rightarrow$

    _____ (j) $CuCl_2$ + Fe $\rightarrow$

    _____ (k) $Cr_2(SO_4)_3$ + K_2CO_3 $\rightarrow$

    _____ (l) C_2H_5OH + O_2 $\rightarrow$

_____ (m) $H_2 + F_2 \rightarrow$

_____ (n) $Ag_2O \rightarrow$

_____ (o) $Cl_2 + KI \rightarrow$

2. For each of the following:

   I. complete the word equation by predicting the products

   II. classify the reaction as synthesis (S), decomposition (D), single replacement (SR), double replacement (DR), or neutralization (N)

   III. write a balanced equation for each word equation

_____ (a) sodium + chlorine $\rightarrow$

_____ (b) gallium fluoride + cesium $\rightarrow$

_____ (c) calcium hydroxide + nitric acid $\rightarrow$

_____ (d) barium chloride + silver nitrate $\rightarrow$

_____ (e) cobalt(II) bromide $\rightarrow$

_____ (f) copper(II) iodide + bromine $\rightarrow$

_____ (g) phosphoric acid + magnesium hydroxide $\rightarrow$

_____ (h) zinc + iodine $\rightarrow$

_____ (i) beryllium chloride $\rightarrow$

_____ (j) iron(III) sulphate + calcium hydroxide $\rightarrow$

Use with textbook pages 256–267.

Types of chemical reactions

Match each Chemical Equation to a Reaction Type below. Each Reaction Type may be used only once.
Chemical Equation
1. _____ $2 KClO_3 \rightarrow 2 KCl + 3 O_2$
2. _____ $16 Al + 3 S_8 \rightarrow 8 Al_2S_3$
3. _____ $LiOH + HNO_3 \rightarrow H_2O + LiNO_3$
4. _____ $2 C_6H_{14} + 19 O_2 \rightarrow 14 H_2O + 12 CO_2$
5. _____ $2 AgNO_3 + Cu \rightarrow Cu(NO_3)_2 + 2 Ag$
6. _____ $Pb(NO_3)_2 + K_2CrO_4 \rightarrow PbCrO_4 + 2 KNO_3$
Reaction Type
A. synthesis
B. combustion
C. neutralization
D. decomposition
E. single replacement
F. double replacement

7. What type of chemical reaction involves two smaller molecules reacting to produce one larger molecule?

  A. synthesis

  B. combustion

  C. decomposition

  D. single replacement

8. Carbon dioxide gas can be broken down into solid carbon and oxygen gas. What type of reaction is this?

  A. synthesis   **C.** neutralization

  B. combustion   **D.** decomposition

Use the following word equation to answer question 9.

potassium chlorate $\rightarrow$ oxygen + potassium chloride

9. What type of reaction is represented by the word equation?

  A. synthesis

  B. decomposition

  C. single replacement

  D. double replacement

10. Which of the following represents a single replacement reaction?

I.	$Sn + 2 AgNO_3 \rightarrow Sn(NO_3)_2 + 2 Ag$
II.	gold(II) cyanide + zinc → gold + zinc cyanide
III.	Magnesium iodide reacts with bromine gas to produce magnesium bromide and iodine.

  A. I and II only   **C.** II and III only

  B. I and III only   **D.** I, II, and III

11. Which set of ordered coefficients balances the following equation?

$$Fe + O_2 \rightarrow Fe_2O_3$$

  A. 2, 1, 1   **C.** 4, 2, 3

  B. 2, 2, 2   **D.** 4, 3, 2

12. What coefficient is needed for water in order to balance the following equation?

$$C_2H_6 + O_2 \rightarrow CO_2 + H_2O$$

  A. 2   **C.** 4

  B. 3   **D.** 6

13. Hydrochloric acid can be used to neutralize barium hydroxide. What is the formula for the salt produced by this neutralization?

  A. $BaCl_2$ **C.** $Ba(ClO_2)_2$

  B. $Ba(ClO)_2$ **D.** $Ba(ClO_3)_2$

14. Which reactants form the salt $MgSO_4$ in a neutralization reaction?

A. SO_2 and MgO_2

B. H_2S and $MgOH$

C. H_2O and $Mg(OH)_2$

D. H_2SO_4 and $Mg(OH)_2$

15. Given the incomplete equation of a chemical reaction: $C_9H_6O_4 + O_2 \rightarrow$

Which of the following are the products formed from this reaction?

I.	H_2
II.	H_2O
III.	CO_2

A. I and II only

B. I and III only

C. II and III only

D. I, II, and III

16. Given the incomplete equation of a chemical reaction:

barium chloride + ammonium carbonate $\rightarrow$

Which of the following are the products formed from this reaction?

I.	H_2O
II.	NH_4Cl
III.	$BaCO_3$

A. I and II only

B. I and III only

C. II and III only

D. I, II, and III

Use the following chemical reaction to answer question 17.

$$2\ HNO_3 + Be(OH)_2 \rightarrow Be(NO_3)_2 + 2\ H_2O$$

17. Which of the following statements is true?

I.	HNO_3 is an acid.
II.	$Be(NO_3)_2$ is a base.
III.	This is a neutralization reaction.
IV.	The products of this reaction are a salt and water.

A. I, II, and III only

B. I, II, and IV only

C. I, III, and IV only

D. II, III, and IV only

18. Sodium nitrate is produced as a result of mixing a solution of cadmium(II) nitrate with a solution of sodium sulphide. What is the other compound formed from this reaction?

A. CdS

B. $CdSO_4$

C. NaS_2

D. $CdNO_4$

Factors Affecting the Rate of Chemical Reactions

Textbook pages 272–281

Before You Read

What do you already know about the speed of chemical reactions? Outline your ideas in the lines below.

_____

_____

What is rate of reaction and how does it apply to chemical reactions?

In a chemical reaction, how quickly or slowly reactants turn into products is called the **rate of reaction**. A reaction that takes a long time has a low reaction rate. A reaction that occurs quickly has a high reaction rate. A *rate* describes how quickly or slowly a change occurs. Every chemical reaction proceeds at a definite rate. However, you can speed up or slow down the rate of a chemical reaction.

What factors affect the rate of a chemical reaction?

The four main factors that affect the rate of chemical reactions are temperature, concentration, surface area, and the presence of a catalyst.

1. Increasing the **temperature** causes the particles (atoms or molecules) of the reactants to move more quickly so that they collide with each other more frequently and with more energy. Thus, the higher the temperature, the greater the rate of reaction. If you decrease the temperature, the opposite effect occurs. The particles move more slowly, colliding less frequently and with less energy. In this case, the rate of reaction decreases. ✔

2. Concentration refers to how much solute is dissolved in a solution.
If a greater concentration of reactant atoms and molecules is present, there is a greater chance that collisions will occur among them. More collisions mean a higher reaction rate. Thus, increasing the concentration of the reactants usually results in a higher reaction rate. At lower concentrations, there is less chance for collisions between particles. This

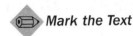
Mark the Text

Reinforce Your Understanding

As you read the section, highlight the main point of each paragraph. Then write out an example that helps you explain this main point.

✔ **Reading Check**

How does temperature affect the rate of a chemical reaction?

_____

_____

_____

_____

means that decreasing the concentrations of the reactants results in a lower reaction rate.

3. Surface area is the measure of how much area of an object is exposed.

For the same mass, many small particles have a greater total surface area than one large particle. For example, steel wool has a larger surface area than a block of steel of the same mass. This allows oxygen molecules to collide with many more iron atoms per unit of time. The more surface contact between reactants, the higher the rate of reaction. The less surface contact, the lower the reaction rate.

Surface area can also be important if a reaction occurs between two liquids that do not mix. In this case, the reaction occurs only at the boundary where the two liquids meet. It is also important to note that not all reactions depend on surface area. If both reactants are gases or liquids that mix together, then there is no surface, and surface area is not a factor.

4. A **catalyst** is a substance that speeds up the rate of a chemical reaction without being used up in the reaction itself. Catalysts reduce the amount of energy required to break and form bonds during a chemical reaction. When catalysts are used, a reaction can proceed although less energy is added during the reaction. For example, enzymes are catalysts that allow chemical reactions to occur at relatively low temperatures within the body.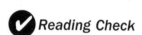

 Reading Check

Is a catalyst used up in a chemical reaction?

_____

Use with textbook pages 272–277.

Rate of chemical reactions

Vocabulary	
catalyst	energy
catalytic converter	heat
collisions	rate of reaction
concentration	surface area
dilute	temperature

Use the terms in the vocabulary box to fill in the blanks. You may use each term only once.

1. A freshly exposed surface of metallic sodium tarnishes almost instantly if exposed to air and moisture, while iron will slowly turn to rust under the same conditions. In these two situations, the _____ refers to how quickly or slowly reactants turn into products.

2. Adding _____ will increase the rate of reaction because this causes the particles of the reactants to move more quickly, resulting in more collisions and more _____ .

3. Removing heat will lower the _____, causing the particles of the reactants to slow down, resulting in less frequent collisions.

4. _____ refers to how much solute is dissolved in a solution. If there is a greater concentration of reactant particles present, there is a greater chance that _____ among them will occur. More collisions mean a higher rate of reaction.

5. A concentrated acid solution will react more quickly than a _____ acid solution because there are more molecules present, increasing the chance of collisions.

6. Grains of sugar have a greater _____ than a solid cube of sugar of the same mass, and therefore will dissolve quicker in water.

7. A _____ , for example an enzyme, is used to speed up a chemical reaction but is not used up in the reaction itself.

8. A _____ in a car has metallic catalysts where several reactions occur. Carbon monoxide, which was produced in the combustion of gasoline, is changed into carbon dioxide and water in the presence of these metallic catalysts.

Use with textbook pages 272–277.

Different rates of reactions

1. Indicate whether each of the following would increase or decrease the rate of reaction.

(a) adding heat _____

(b) removing heat _____

(c) adding a catalyst _____

(d) diluting a solution _____

(e) removing an enzyme _____

(f) lowering the temperature _____

(g) increasing the temperature _____

(h) decreasing the surface area _____

(i) increasing the concentration of a solution

_____

(j) breaking a reactant down into smaller pieces

_____

2. Identify which situation would have a higher reaction rate. Then state the factor that affected the rate of reaction in each situation.

	Situation X	Situation Y	Situation with a higher reaction rate (X or Y)	Factor affecting the rate of reaction
(a)	1 g of sugar (cubes) 	1 g of sugar (grains) 		

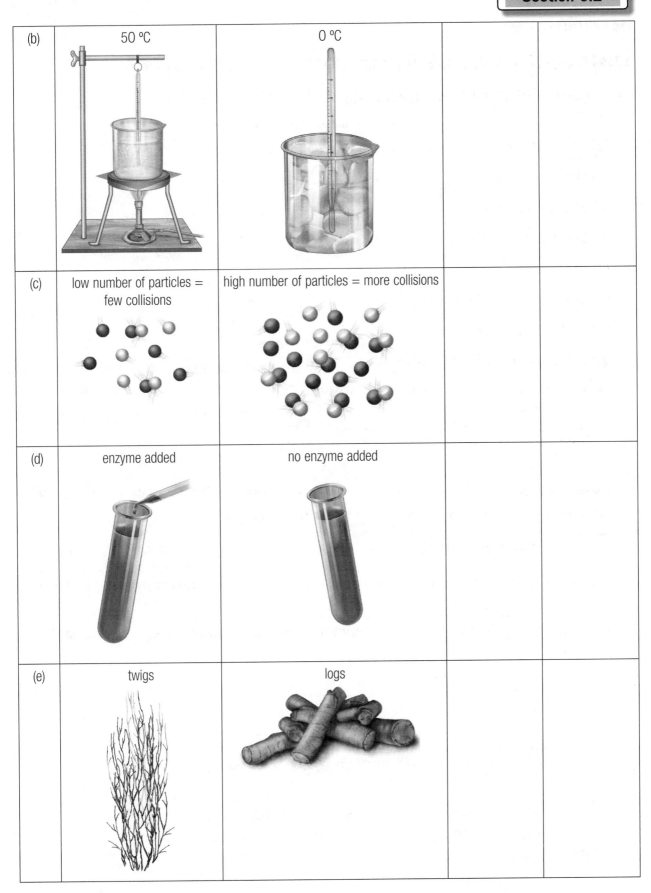

(b)   50 °C   0 °C

(c)   low number of particles =   high number of particles = more collisions
       few collisions

(d)   enzyme added   no enzyme added

(e)   twigs   logs

Use with textbook pages 272–277.

Four factors affecting the rate of reactions

Use the following graph to answer question 1.

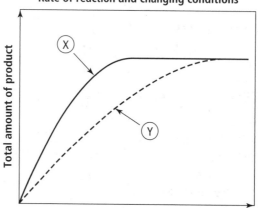

Rate of reaction and changing conditions

Total amount of product (y-axis)

Time from start of reaction (x-axis)

1. The graph above shows the differences in the rate of reaction at different temperatures, concentrations, surface area, and the presence or absence of a catalyst. A steeper line represents a greater rate of reaction. Indicate which line (X or Y) each of the following are associated with.

(a) lower temperature _____   (b) higher temperature _____

(c) lower concentration _____   (d) higher concentration _____

(e) absence of a catalyst _____   (f) presence of a catalyst _____

(g) larger pieces (small surface area) _____

(h) smaller pieces (large surface area) _____

2. Which of the four factors affecting reaction rate is most important in each of the following examples? Choose from concentration, temperature, surface area, and catalyst.

(a) Raw carrots are cut into thin slices for cooking. _____

(b) Protein is broken down in the stomach by the enzyme pepsin. _____

(c) A woolly mammoth is found, perfectly preserved, near the Arctic. _____

(d) More bubbles appear when a concentrated solution of hydrochloric acid is added to a magnesium strip than when a dilute solution of the acid is added. _____

Use with textbook pages 272–277.

Factors affecting the rate of chemical reactions

Match the Term on the left with the best Descriptor on the right. Each Descriptor may be used only once.

Term	Descriptor
1. _____ catalyst **2.** _____ temperature **3.** _____ surface area **4.** _____ concentration **5.** _____ rate of reaction **6.** _____ catalytic converter	**A.** a measure of how much area of an object is exposed **B.** the amount of substance dissolved in a given volume of solution **C.** a measure of the average kinetic energy of all the particles in a sample of matter **D.** a substance that speeds up the rate of a chemical reaction without being used up itself or changed **E.** a measure of how quickly products form, or given amounts of reactants react, in a chemical reaction **F.** a stainless steel pollution-control device that converts poisonous gases from the vehicle's exhaust into less harmful substances

7. When you walk through a crowded hallway at school, you are more likely to bump into another person. To which of the following factors that affect rate of reaction is this analogy referring?

A. catalyst   **C.** surface area

B. temperature   **D.** concentration

8. Which of the following are true about how temperature affects the rate of reaction?

I.	heating causes the particles of the reactants to move more quickly
II.	lowering the temperature will raise the energy level of the particles
III.	increasing the temperature results in more collisions between the particles

A. I and II only

B. I and III only

C. II and III only

D. I, II, and III

9. Increasing which of the following will increase the frequency of collisions?

I.	temperature
II.	surface area
III.	concentration

A. I and II only

B. I and III only

C. II and III only

D. I, II, and III

10. Which of the following will lower the rate of reaction?

A. adding an enzyme to the reaction

B. decreasing the temperature from 40°C to 10°C

C. breaking a chunk of calcium up into smaller pieces

D. increasing the amount of solute dissolved in a solution

Atomic Theory, Isotopes, and Radioactive Decay

Textbook pages 286–301

Before You Read

Radiation is used for many purposes. What uses of radiation are you already aware of? Write your response in the lines below.

? Create a Quiz

After you have read this section, create a five-question quiz based on what you have learned. Answer the questions until you get each one correct.

What is radioactivity?

Radioactivity is the release of high-energy particles and rays of energy from a substance as a result of changes in the nuclei of its atoms. **Radiation** refers to high-energy rays and particles emitted by radioactive sources, including radio waves, microwaves, infrared rays, visible light, and ultraviolet rays, that are found on the electromagnetic spectrum. **Light** is a form of radiation that humans can see.

What are isotopes?

Isotopes are different atoms of a particular element that have the same number of protons but different numbers of neutrons. The **mass number** of an atom is an integer (whole number) that represents the sum of the atom's protons and neutrons—so isotopes have different mass numbers. The mass number of an isotope is found by adding the atomic number (number of protons) to the number of neutrons.

Mass number = atomic number + number of neutrons

To find the number of neutrons of an isotope, subtract the atomic number from the mass number.

✔ Reading Check

Write the equation used to calculate mass number.

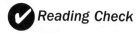

Number of neutrons = mass number – atomic number

How are isotopes represented?

Chemists represent isotopes using standard atomic notation (also called the **nuclear symbol**), a shortened form involving the chemical symbol, atomic number, and mass number. The mass number is written as a superscript (above) on the left of the symbol. The atomic number is written as a subscript (below), also on the left.

$$_{19}^{39}\text{K}$$

The mass number of this potassium isotope is 39. The atomic number is 19. An isotope of potassium with a mass number of 39 can also be represented as potassium-39, or K-39

What is radioactive decay?

By emitting radiation, atoms of one kind of element can change into atoms of another element. Radioactive atoms emit radiation because their nuclei are unstable. Unstable atoms gain stability by losing energy. **Radioactive decay** is the process in which unstable nuclei lose energy by emitting radiation. Unstable radioactive atoms undergo radioactive decay and form stable, non-radioactive atoms, usually of a different element. **Radioisotopes** are natural or human-made isotopes that decay into other isotopes, releasing radiation.

What different types of radiation are emitted during radioactive decay?

The three major types of radiation are alpha radiation, beta radiation, and gamma radiation. Their properties are summed up in the following table: ✔

✔ Reading Check

Name the three main types of radiation.

_____

_____

_____

Table 7.3 Properties of Alpha, Beta, and Gamma Radiation

Property	Alpha Radiation	Beta Radiation	Gamma Radiation
Symbol	$_{2}^{4}\text{a}$ or $_{2}^{4}\text{He}$	$_{-1}^{0}\text{b}$ or $_{-1}^{0}e$	$_{0}^{0}\text{c}$
Composition	Alpha particles	Beta particles	High-energy electromagnetic radiation
Description of radiation	Helium nuclei, $_{2}^{4}\text{He}$	Electrons	High energy rays
Charge	2+	1−	0
Relative penetrating power	Blocked by paper	Blocked by metal foil or concrete	Partly or completely blocked by lead

How is radioactive decay expressed?

Radioactivity results when the nucleus of an atom decays.
There are three radioactive decay processes:

1. Alpha decay: The emission of an **alpha particle** (the
    same particles found in the nucleus of a helium atom)
    from a nucleus is a process called **alpha decay.** When
    a radioactive nucleus emits an alpha particle, the atomic
    number of the product nucleus is reduced by two, and its
    mass number by four. However, the sum of the atomic
    numbers and the sum of the mass numbers on each side of
    the arrow remain equal.

$$^{226}_{88}\text{Ra} \rightarrow {}^{222}_{86}\text{Rn} + {}^{4}_{2}\alpha$$

2. Beta decay: In **beta decay**, a neutron changes into a
    proton and a **beta particle**, an electron. The proton
    remains in the nucleus while the electron leaves the
    nucleus. Since the proton remains in the nucleus, the
    atomic number of the element increases by one—it
    has become an atom of the next higher element on the
    periodic table. However, its mass number does not
    change, as a proton of almost equal mass has replaced the
    neutron.

$$^{131}_{53}\text{I} \rightarrow {}^{131}_{54}\text{Xe} + {}^{0}_{-1}\beta$$

3. Gamma decay: **Gamma decay** results from a
    redistribution of energy within the nucleus. **Gamma
    radiation** consists of rays of high-energy, short-
    wavelength radiation. A gamma ray is given off as the
    isotope changes from a high-energy state to a lower
    energy state.

$$^{60}_{28}\text{Ni*} \rightarrow {}^{60}_{28}\text{Ni} + {}^{0}_{0}\gamma$$

The "*" means that the nickel nucleus has extra energy that is
released as a gamma ray.

Use with textbook pages 289–293.

Isotopes

1. What is an isotope?

_____

2. Atomic number + number of neutrons = _____

3. Number of protons + number of neutrons = _____

4. Mass number – atomic number = _____

Use the following standard atomic notation of an isotope to answer questions 5 to 7.

5. Label the mass number and the atomic number.

6. What is the name of this isotope? _____

7. Determine the number of subatomic particles for this isotope:

    (a) number of protons = _____

    (b) number of electrons = _____

    (c) number of neutrons = _____

8. In each of the following cases, what element does the symbol X represent and how many neutrons are in the nucleus?

    (a) $^{21}_{10}$ X   Element = _____

                                     Number of neutrons = _____

    (b) $^{32}_{16}$ X   Element = _____

                                     Number of neutrons = _____

    (c) $^{230}_{89}$ X   Element = _____

                                     Number of neutrons = _____

    (d) $^{234}_{90}$ X   Element = _____

                                     Number of neutrons = _____

9. Complete the following table. The first row has been completed to help guide you.

Isotope	Standard atomic notation	Atomic number	Mass number	Number of protons	Number of neutrons
carbon-14	$^{14}_{6}\text{C}$	6	14	6	8
		27	52		
nickel-60					
			1 4	7	
thallium-201					
	$^{226}_{88}\text{Ra}$				
				82	126

Use with textbook pages 294–297.

Alpha, beta, and gamma radiation

1. Label the following diagram. Identify the penetrating power of the three forms of radioactive decay products: alpha particle, beta particle, and gamma ray.

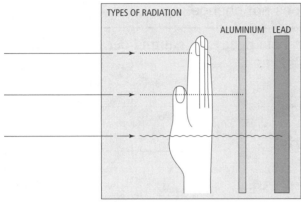

2. Indicate whether the description is referring to an alpha particle, a beta particle, or a gamma ray. The description can refer to more than one of the forms of radiation.

(a) $_0^0 \gamma$ _____

(b) $_{-1}^0 \beta$ or $_{-1}^0 e$ _____

(c) $_2^4 \alpha$ $_2^4 He$ _____

(d) has a charge of 0 _____

(e) has a charge of 1- _____

(f) has a charge of 2+ _____

(g) is a helium nucleus _____

(h) is a high-speed electron _____

(i) is emitted from the nucleus _____

(j) is emitted only during beta decay _____

(k) is emitted only during alpha decay _____

(l) can be stopped by aluminum foil _____

(m) is emitted only during gamma decay _____

(n) is affected by electric and magnetic fields _____

(o) is not affected by electric and magnetic fields _____

(p) is a high energy wave with short wavelengths _____

(q) is the highest energy form of electromagnetic radiation _____

(r) has low penetrating power (can be stopped by a single piece of paper) _____

(s) has the greatest penetrating power (can only be stopped by lead or concrete) _____

Use with textbook pages 286–299.

Radioactive decay and nuclear equations

Remember the following two rules when working with nuclear equations:

I. The sum of the mass numbers does not change.

II. The sum of the charges in the nucleus does not change.

Identify each nuclear equation as alpha decay, beta decay, or gamma decay, and then complete the nuclear equation.

1. $^{32}_{15}P$ ---> $^{32}_{16}S$ + _____ _____

2. $^{218}_{84}Po$ ---> _____ + $^{4}_{2}He$ _____

3. _____ ---> $^{35}_{18}Ar$ + $^{0}_{-1}e$ _____

4. $^{24}_{12}Mg^{*}$ ---> _____ + $^{0}_{0}\gamma$ _____

5. $^{234}_{91}Pa$ ---> _____ + $^{4}_{2}\alpha$ _____

6. $^{141}_{58}Ce$ ---> _____ + $^{0}_{-1}e$ _____

7. $^{216}_{84}Po$ ---> _____ + $^{0}_{-1}\beta$ _____

8. $^{20}_{9}F$ ---> $^{20}_{10}Ne$ + _____ _____

9. $^{58}_{26}Fe^{*}$ ---> $^{58}_{26}Fe$ + _____ _____

10. _____ ---> $^{221}_{87}Fr$ + $^{4}_{2}\alpha$ _____

11. $^{149}_{64}Gd^{*}$ ---> _____ + $^{0}_{0}\gamma$ _____

12. $^{226}_{88}Ra$ ---> $^{222}_{86}Rn$ + _____ _____

13. _____ ---> $^{212}_{82}Pb$ + $^{0}_{-1}\beta$ _____

14. $^{214}_{83}Bi$ ---> $^{210}_{81}Tl$ + _____ _____

15. _____ ---> $^{254}_{98}Cf$ + $^{0}_{0}\gamma$ _____

Use with textbook pages 286–299.

Atomic theory, isotopes, and radioactive decay

Match the Descriptor on the left with the best Scientist on the right. Each Scientist may be used more than once.	
Descriptor	**Scientist**
1. _____ discovered X-rays 2. _____ identified polonium and radium 3. _____ first to identify alpha, beta, and gamma radiation 4. _____ discovered the nucleus and created a model of the atom 5. _____ discovered that uranium salts emitted rays that darkened photographic plates	**A.** Marie Curie **B.** Henri Becquerel **C.** Ernest Rutherford **D.** Wilhelm Roentgen

6. Which of the following electromagnetic radiations has the highest frequency and energy?

A. X-rays

B. gamma rays

C. microwaves

D. ultraviolet radiation

7. The number of neutrons in an atom is found by

A. adding the atomic number to the mass number

B. subtracting the mass number from the atomic number

C subtracting the atomic number from the mass number

D. adding the number of protons to the number of electrons

8. What is used to tell different isotopes of a particular element apart?

A. the mass number

B. the atomic number

C. the number of protons

D. the number of electrons

9. One isotope of polonium is $^{212}_{84}$Po. Any other isotope of polonium must have

A. 84 protons

B. 128 protons

C. 84 neutrons

D. 128 neutrons

10. How many protons, neutrons, and electrons are in the isotope calcium-42, $^{42}_{20}$Ca?

	Protons	**Neutrons**	**Electrons**
A.	20	22	20
B.	20	20	22
C.	22	22	20
D.	22	20	20

Use the following standard atomic notation for the lithium isotope to answer question 11.

$$^{7}_{3}\text{Li}$$

11. What does each part of the standard atomic notation shown above represent?

	"3"	**"7"**
A.	atomic number	mass number
B.	mass number	atomic number
C.	number of neutrons	number of protons
D.	number of protons	number of electrons

Use the following diagram showing the penetrating power of a type of radiation to answer question 12.

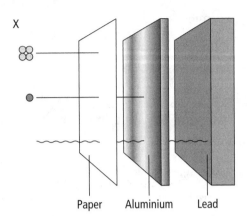

X

Paper   Aluminium   Lead

12. What does "X" represent?

 A. a gamma ray

 B. a beta particle

 C. an alpha particle

 D. a high-speed electron

13. Which type of radioactive decay process results in no change to the nucleus?

 A. beta decay

 B. alpha decay

 C. gamma decay

14. The symbol $^{4}_{2}\text{He}$ is equivalent to which of the following?

 A. $^{0}_{-1}e$

 B. $^{0}_{-1}\beta$

 C. $^{0}_{0}\gamma$

 D. $^{4}_{2}\alpha$

15. Which of the following represents a beta decay?

 A. $^{131}_{53}\text{I} \rightarrow ^{131}_{54}\text{Xe} + ^{0}_{-1}e$

 B. $^{60}_{28}\text{Ni*} \rightarrow ^{60}_{28}\text{Ni} + ^{0}_{0}\gamma$

 C. $^{226}_{88}\text{Ra} \rightarrow ^{222}_{86}\text{Rn} + ^{4}_{2}\alpha$

 D. $^{231}_{91}\text{Pa} \rightarrow ^{227}_{89}\text{Ac} + ^{4}_{2}\text{He}$

Use the following incomplete nuclear equation to answer question 16.

$$^{144}_{60}\text{NdI} \rightarrow \underline{\quad} + ^{4}_{2}\alpha$$

16. What is product of this decay process?

 A cobalt-58

 B. cerium-58

 C. cerium-140

 D. samarium-62

Half-Life

Textbook pages 302–311

Before You Read

Write a sentence in the lines below explaining what the word decay means to you. As you read about radioactive decay, think about how the common meaning of decay differs from the scientific meaning.

_____

_____

What is radiocarbon dating?

Radiocarbon dating is the process of determining the age of an object by measuring the amount of carbon-14 remaining in that object. Carbon's isotopes include carbon-12 and carbon-14. When an organism is alive, the ratio of carbon-14 atoms to carbon-12 atoms in the organism remains nearly constant. But when an organism dies, its carbon-14 atoms decay without being replaced. The ratio of carbon-14 to carbon-12 then decreases with time. By measuring this ratio, the age of an organism's remains can be estimated. Only material from plants and animals that lived within the past 50 000 years contains enough carbon-14 to be measured using radiocarbon dating. ✔

What is a half-life and how is it used in radiocarbon dating?

A half-life is a measure of the rate of radioactive decay for a given isotope. It is equal to the time required for half the nuclei in a sample to decay. Its value is a constant for any radioactive isotope. For example, the half-life of the radioisotope strontium-90 is 29 years. If you have 10.0 g of strontium-90 today, 29 years from now you will have 5.00 g left. This is because one half-life will have passed $(10.0 \text{ g} \times \frac{1}{2} = 5.00 \text{ g})$. 58 years from now, two half-lives will have passed and 2.50 g of the sample will remain $(10.0 \text{ g} \times \frac{1}{2} \times \frac{1}{2} = 2.50 \text{ g})$. The shorter the half-life is, the faster the decay rate. A **decay curve** is a curved line on a graph that shows the rate at which radioisotopes decay.

 Mark the Text

In Your Own Words

After you read this section, go back and summarize the main concepts in your own words.

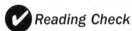 Reading Check

1. Which carbon isotope undergoes radioactive decay?

_____

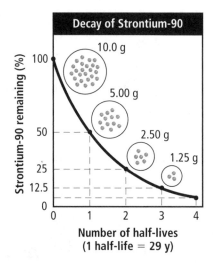

Decay of Strontium-90

This decay curve shows how the amount of strontium-90 in a sample changes over time.

What are parent and daughter isotopes?

A **parent isotope** is an isotope that undergoes radioactive decay. The stable product of this decay is called the **daughter isotope**. The production of a daughter isotope can be a direct reaction or the result of a series of decays.

Each parent isotope can be paired with a specific daughter isotope. For example, carbon-12 is the daughter isotope of carbon-14 (the parent isotope). The chart on page 307 of the textbook lists other common isotope pairs. It also shows the half-life of the parent and the effective dating range the isotope can be used for. ✔

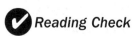

Reading Check

1. Which isotope decays, the parent or the daughter?

How does the potassium-40 clock work?

The potassium-40 clock uses radioisotopes, specifically potassium-40 and argon-40, to determine Earth's age. Potassium-40 has a half-life of 1.3 billion years. Its daughter isotope is argon-40. When rock is produced from lava, all the gases in the molten rock, including argon-40, are driven out. This process sets the potassium radioisotope clock to zero, because potassium-40 (the parent) is present but no argon-40 (the daughter) is present.

As the molten rock cools over time, it traps gases that form as a result of radioactive decay. When tested, both potassium-40 and argon-40 are now present in the rock. As the mass of the parent isotope drops, the mass of the daughter isotope increases. By measuring this ratio, the age of the rock can be estimated. For example, if analysis showed that there were equal masses of potassium-40 and argon-40 in a rock, the rock would be 1.3 billion years old, the amount of time it takes half of the potassium-40 to decay into argon-40.

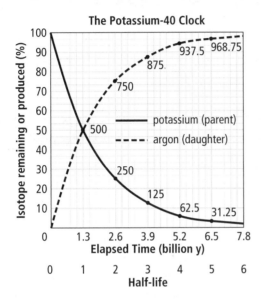

The solid line shows that the parent isotope is decaying. The dashed line shows that the daughter isotope is being produced.

Use with textbook pages 302–309.

Radioactive decay

1. Define the following terms.

(a) half-life _____

(b) decay curve _____

(c) parent isotope _____

(d) daughter isotope _____

2. Complete the following tables.

Half-Life	Percent of parent isotope	Percent of daughter isotope
0		
1		
2		
3		
4		

Half-Life	Fraction of parent isotope	Fraction of daughter isotope
0		
1		
2		
3		
4		

3. A rock sample contains 120 g of a radioactive isotope. The radioactive isotope has a half-life of 5 years.

(a) Complete the following table.

Half-Life	Time (a)	Mass (g)
0	0	
1	5	
2	10	
3	15	
4	20	
5	25	

(b) How much of the radioactive isotope if left after 25 years have passed? _____

(c) How many half-lives have passed if there is only 15 g of the parent isotope left?

_____

(d) How many years have passed if there is only 7.5 g of the parent isotope left?

_____

(e) Use the data in the table to graph a decay curve. Label the *x*-axis with Time (a) and the *y*-axis with Mass (g).

4. A rock sample contains 80 g of a radioactive isotope with a half-life of 20 years.

(a) Complete the following table.

Half-Life	Time (a)	Mass of parent isotope (g)	Mass of daughter isotope (g)
0	0		
1	20		
2	40		
3	60		
4	80		
5	100		

(b) How much of the parent isotope is left after 4 half-lives? _____

(c) How much of the parent isotope is left after 100 years? _____

(d) How much of the daughter isotope is present after 60 years? _____

(e) How much time has passed if 77.5 g of the daughter isotope is present? _____

(f) What is the ratio of parent isotope to daughter isotope after 2 half-lives? _____

Use with textbook pages 302–309.

Calculating half-life

1. A radioactive isotope has a half-life of 10 minutes.

(a) What fraction of the parent isotope will be left after 30 minutes?

_____

(b) What percent of the parent isotope will be left after 40 minutes?

_____

(c) What fraction of the daughter isotope will be present after 20 minutes?

_____

(d) What percent of the daughter isotope will be present after 50 minutes?

_____

2. A 36 g sample of a radioactive isotope decayed to 4.5 g in 36 minutes. How much of the original parent isotope would remain after the first 12 minutes?

_____

3. The half-life of a particular radioactive isotope is 8 hours. What percent of the parent isotope would remain after 1 day? _____

4. A radioactive isotope sample has a half-life of 4 days. If 6 g of the sample remains unchanged after 12 days, what was the initial mass of the sample?

_____

5. Suppose the ratio of a radioactive parent isotope to a stable daughter isotope within a rock sample is 1:3. The half-life of the parent isotope is 710 million years. How old is the rock sample? _____

6. A rock sample was dated using potassium-40. Measurement indicates that 1/8 of the original parent isotope is left in the rock sample. How old is the rock sample?

_____

7. When a sample of lava solidified, it contained 28 g of uranium-238. If that lava sample was later found to contain only 7 g of U-238, how many years had passed since the lava solidified? _____

8. After 25 years, the number of radioactive cobalt atoms in a sample is reduced to $\frac{1}{32}$ of the original count. What is the half-life of this isotope? _____

9. The half-life of Sr-90 is 28 years. If an 80 g sample of Sr-90 is currently in a sample of soil, how much Sr-90 will be present in the soil 84 years later? _____

Use with textbook pages 305–309.

Decay curves

1. Use the decay curve to answer the questions.

(a) What is the half-life of the isotope?

_____

(b) How much of the parent isotope remains after 4 days? _____

(c) How much of the daughter isotope is present after 6 days? _____

(d) What fraction of the parent isotope remains after 8 days? _____

(e) How long does it take for the parent isotope to decay to 5 g? _____

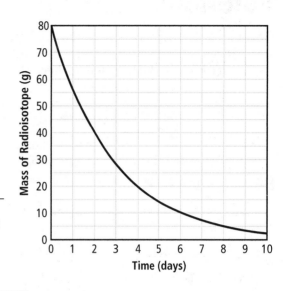

2. Use the decay curve to answer the questions.

(a) What is the common isotope pair for this decay curve? _____

(b) What is the half-life of the parent isotope?

_____

(c) What does the intersection of the two lines represent? _____

(d) What fraction of the daughter isotope is present after 5.2 billion years have passed?

_____

(e) What is the ratio of parent isotope to daughter isotope after 2.6 billion years have passed? _____

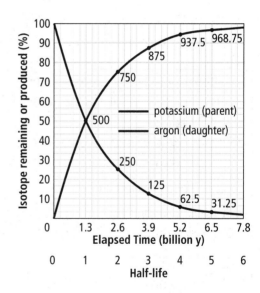

Use with textbook pages 302–309.

Half-life

Match the Term on the left with the best Descriptor on the right. Each Descriptor may be used only once.	
Term	**Descriptor**
1. _____ half-life 2. _____ decay curve 3. _____ parent isotope 4. _____ daughter isotope	**A.** the stable product of radioactive decay **B.** the isotope that undergoes radioactive decay **C.** a curved line on a graph that shows the rate at which radioisotopes decay **D.** the time required for half the nuclei in a sample of a radioactive isotope to decay

5. Radiocarbon dating can be used to determine the age of which of the following?

I.	a rock sample
II.	the fossil of a fern plant
III.	the skeleton of a dead bear

A. I and II only

B. I and III only

C. II and III only

D. I, II, and III

6. After how many half-lives are there equal amounts of parent and daughter isotopes?

A. 1   **C.** 3

B. 2   **D.** 4

7. The half-life of Ni-28 is six days. What fraction of a sample of this isotope will remain after 18 days?

A. 1/2   **C.** 1/8

B. 1/4   **D.** 1/16

8. The half-life of a particular radioactive isotope is 6 hours. What percent of the daughter isotope would be present after 1 day?

A. 50%   **C.** 87.5%

B. 75%   **D.** 93.75%

9. A 24 g sample of a radioactive isotope decayed to 1.5 g in 48 minutes. How much of the original parent isotope remained after 24 minutes?

A. 3 g   **C.** 12 g

B. 6 g   **D.** 18 g

10. A radioactive isotope sample has a half-life of 5 days. If 8 g of the sample remains unchanged after 20 days, what was the initial mass of the sample?

A. 32 g   **C.** 128 g

B. 64 g   **D.** 256 g

11. If the half-life of an isotope is 8000 years and the amount of that isotope present in an igneous rock is only $\frac{1}{4}$ of the original amount, how old is the rock?

A. 8000 years old

B. 16 000 years old

C. 24 000 years old

D. 32 000 years old

12. What is the advantage of using a radioisotope with a short half-life for medical diagnostic purposes?

A. the radioactivity is easy to monitor

B. the radioactivity lasts for a long time

C. the radioactivity does not stay in the body

D. the radioactivity induced by the radioisotope is stronger

Nuclear Reactions

Textbook pages 312–325

Before You Read

Nuclear reactors supply energy to many parts of Canada. Summarize what you already know about nuclear reactions in the lines below.

_____

_____

What is nuclear fission?

Nuclear fission is a nuclear reaction in which a nucleus breaks apart, producing two or more smaller nuclei, subatomic particles, and energy. For example, for Uranium-235,

$$^{1}_{0}n + ^{235}_{92}U \longrightarrow ^{92}_{36}Kr + ^{141}_{56}Ba + 3\,^{1}_{0}n + energy$$

Heavy nuclei tend to be unstable because of the repulsive forces between their many protons. In order to increase their stability, atoms with heavy nuclei may split into atoms with lighter nuclei. The fission of a nucleus is accompanied by a very large release of energy. Fission is the source of energy for all nuclear power generation used today; however, the radioactive daughter products are a significant waste disposal problem. ✔

How do nuclear reactions work?

In typical chemical reactions, the energy produced or used is so small that there is very little change in mass. There are no changes in the nuclei of the reactants, so the identities of the atoms do not change. Chemical reactions involve electrons and rearrangements in the way atoms and ions are connected to each other.

Mark the Text

Summarize

As you read this section, highlight the main point in each paragraph. Then write a short paragraph summarizing what you have learned.

✔ **Reading Check**

Why do heavy nuclei tend to be unstable?

_____

_____

_____

A **nuclear reaction** is a process in which an atom's nucleus changes by gaining or releasing particles or energy. A nuclear reaction can release protons, neutrons, and electrons, as well as gamma rays. In nuclear reactions, a small change in mass results in a very large change in energy.

Scientists can *induce,* or cause, a nuclear reaction by making a nucleus unstable, causing it to undergo a reaction immediately. Bombarding a nucleus with alpha particles, beta particles, or gamma rays induces a nuclear reaction. An example of an induced reaction is given below. Nitrogen-14 is bombarded with alpha particles, producing oxygen and protons.

$$ {}^{4}_{2}\alpha + {}^{14}_{7}\text{N} \longrightarrow {}^{17}_{8}\text{O} + {}^{1}_{1}p $$

When some nuclei undergo fission, they release subatomic particles that trigger more fission reactions. This ongoing process in which one reaction initiates the next reaction is called a **chain reaction**. The number of fissions and the amount of energy released can increase rapidly and lead to a violent nuclear explosion. Uranium-235, which is used in Canadian nuclear reactors, undergoes such a reaction. Keeping the chain reaction going in a nuclear power plant, while preventing it from racing out of control, requires precise monitoring and continual adjusting.

What are the rules for writing nuclear equations?

A **nuclear equation** is a set of symbols that indicates changes in the nuclei of atoms during a nuclear reaction. The following rules can be used when you write a nuclear equation.

1. The sum of the mass numbers on each side of the equation stays the same.

2. The sum of the charges (represented by atomic numbers) on each side of the equation stays the same.

What is nuclear fusion?

Nuclear fusion is a nuclear reaction in which small nuclei combine to produce a larger nucleus. Other subatomic particles as well as energy are released in this process. Fusion occurs at the core of the Sun and other stars where sufficient pressure and high temperatures cause isotopes of hydrogen to collide with great force. This forces two nuclei of hydrogen to merge into a single nucleus, releasing an enormous amount of energy. The fusion reaction that occurs in the Sun is given below.

$$\ _{1}^{2}\text{H} + \ _{1}^{3}\text{H} \longrightarrow \ _{2}^{4}\text{He} + \ _{0}^{1}n + \text{energy}$$

We do not currently have the technology to extract energy from fusion reactions. One of the difficulties is achieving and containing the high temperatures and pressures required to bring about fusion. ✔

✔ *Reading Check*

Identify the main difference between fission and fusion.

_____

_____

_____

Use with textbook pages 312–321.

Radioactivity

Vocabulary		
CANDU reactor	neutron	proton
chain reaction	nuclear fission	subatomic particles
energy	nuclear fusion	Sun
induced	nuclear reaction	unstable
isotope		

Use the terms in the vocabulary box to fill in the blanks. You may use each term only once.

1. _____ is the splitting of a heavy nucleus into two lighter nuclei.

2. Heavy nuclei, like those of uranium-238, tend to be _____ due to the repulsive forces between the many protons.

3. Nuclear fission is usually accompanied by a very large release of _____.

4. A _____ occurs when an atom's nucleus changes by gaining or releasing particles or energy. Atoms are changed from one _____ into another, producing different elements.

5. In a nuclear reaction, _____, (e.g. protons, neutrons, and electrons) and gamma rays, can be emitted from the nucleus.

6. A nuclear reaction is _____ by bombarding a nucleus with alpha particles, beta particles, or gamma rays.

7. A _____, ^1_1p, is the same thing as a hydrogen-1 nucleus.

8. A _____, ^1_0n, has a charge of 0 and a mass number of 1.

9. A _____ is an ongoing nuclear reaction in which some products go on to cause more nuclear reactions to occur.

10. The Canadian deuterium uranium reactor, _____, is used for nuclear power generation. It is one of the safest nuclear reactors in the world.

11. _____ is the process in which two smaller nuclei join together to make a bigger one. This process occurs at the core of the _____ and other stars.

Use with textbook pages 312–321.

Comparing nuclear fission and fusion

1. Complete the following table.

	Nuclear fission	**Nuclear fusion**
Give a description of the process.		
What is produced as a result of this nuclear process?		
Are the products radioactive?		
What is needed for this nuclear reaction to occur?		
Where does this process occur?		
Give an example of a nuclear equation.		

2. Identify the following diagrams as nuclear fission or nuclear fusion. Label the parent isotope(s), daughter isotope(s), neutron(s), and energy.

(a) _____

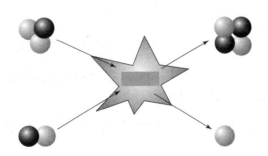

(b) _____

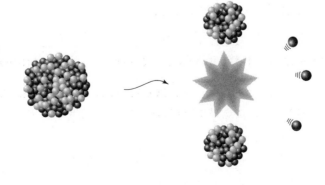

Use with textbook pages 312–321.

Nuclear fission and fusion reactions

Remember the following two rules when working with nuclear equations:

I. The sum of the mass numbers does not change.

II. The sum of the charges in the nucleus does not change.

Identify each nuclear equation (nuclear fission or nuclear fusion) and then complete the nuclear equation.

1. $^{1}_{0}n + {}^{235}_{92}U \longrightarrow {}^{143}_{54}Xe + {}^{90}_{38}Sr + \underline{\quad} {}^{1}_{0}n$

2. $^{2}_{1}H + \underline{\quad} \longrightarrow {}^{1}_{1}H + {}^{3}_{1}H$

3. $^{1}_{0}n + {}^{235}_{92}U \longrightarrow {}^{152}_{60}Nd + \underline{\quad} + 4{}^{1}_{0}n$

4. $^{2}_{1}H + {}^{2}_{1}H \longrightarrow {}^{3}_{2}He + \underline{\quad}$

5. $^{1}_{0}n + \underline{\quad} \longrightarrow {}^{90}_{37}Rb + {}^{143}_{55}Cs + 3{}^{1}_{0}n$

6. $^{2}_{1}H + {}^{3}_{1}H \longrightarrow {}^{4}_{2}He + \underline{\quad}$

7. $^{1}_{0}n + {}^{256}_{100}Fm \longrightarrow \underline{\quad} + {}^{140}_{54}Xe + 4{}^{1}_{0}n$

8. $^{1}_{0}n + {}^{235}_{92}U \longrightarrow {}^{106}_{39}Y + \underline{\quad} + 3{}^{1}_{0}n$

9. $^{1}_{0}n + {}^{235}_{92}U \longrightarrow {}^{115}_{49}In + {}^{118}_{43}Tc + \underline{\quad} {}^{1}_{0}n$

10. $^{1}_{0}n + \underline{\quad} \longrightarrow {}^{137}_{52}Te + {}^{100}_{42}Mo + 3{}^{1}_{0}n$

Use with textbook pages 312–321.

Nuclear reactions

Match each Number on the Diagram of a nuclear reaction on the left with the correct Descriptor on the right. Each Descriptor may be used more than once.

Diagram of a nuclear reaction	Descriptor
	A. energy **B.** neutron **C.** parent isotope **D.** nuclear fusion **E.** nuclear fission **F.** daughter isotope

1. _____
2. _____
3. _____
4. _____
5. _____
6. _____

7. What is the symbol for a proton?

A. $^{4}_{2}\alpha$

B. $^{0}_{0}\gamma$

C. $^{1}_{0}n$

D. $^{1}_{1}p$

8. Which of the following is the source of the Sun's energy?

A. convection

B. nuclear fusion

C. nuclear fission

D. CANDU reactor

9. Which of the following represents a nuclear fusion equation?

A. $^{234}_{90}Th \rightarrow {}^{230}_{88}Ra + {}^{4}_{2}He$

B. $^{238}_{92}U \rightarrow {}^{234}_{90}Th + {}^{4}_{2}He + 2\gamma$

C. $^{2}_{1}H + {}^{3}_{1}H \rightarrow {}^{4}_{2}He + {}^{1}_{0}n + energy$

D. $^{1}_{0}n + {}^{235}_{92}U \rightarrow {}^{92}_{36}Kr + {}^{141}_{56}Ba + 3{}^{1}_{0}n + energy$

10. What is the total mass number?

$$^{92}_{36}Kr + {}^{141}_{56}Ba + 3{}^{1}_{0}n + energy$$

A. 92   **C.** 234

B. 95   **D.** 236

11. How many neutrons are released in this nuclear equation?

$$^{1}_{0}n + {}^{239}_{94}Pu \rightarrow {}^{141}_{54}Xe + {}^{97}_{40}Zr + \mathbf{?}\ {}^{1}_{0}n$$

A. 0   **C.** 2

B. 1   **D.** 3

12. What isotope balances this nuclear reaction?

$$^{1}_{0}n + {}^{235}_{92}U \rightarrow \underline{\quad} + {}^{119}_{50}Sn + 3{}^{1}_{0}n$$

A. $^{114}_{39}Y$ **C.** $^{114}_{42}Mo$

B. $^{117}_{39}Y$ **D.** $^{117}_{42}Mo$

13. What is needed for nuclear fusion to occur?

I.	pressure
II.	a beta particle
III.	high temperature

A. I and II only   **C.** II and III only

B. I and III only   **D.** I, II, and III

The Language of Motion

Textbook pages 344–361

Before You Read

What does the term "uniform" mean to you? If motion is uniform, how does it behave? Write your ideas in the lines below.

_____

_____

Make Flash Cards

Create flash cards for the measurements described in this section. Write a measurement on the front of the card and what it measures on the back. Quiz yourself until you can define each measurement.

Reading Check

1. Measurements of motion can be placed in either of two categories. Name these categories.

_____

_____

How is motion measured?

Motion involves a change in location. There are different ways of measuring motion. These can be placed in two categories:

1. Scalar quantity: A scalar quantity or **scalar** describes the size of a measurement or the amount (number) being counted, a factor known as *magnitude*. A scalar quantity has magnitude only. It does not include direction. Example: You walk 4 km/h.

2. Vector quantity: A vector quantity or **vector** has both magnitude and direction. Example: You walk 4 km/h [E]. ✔

The table below summarizes some of the measurements used to describe motion.

Measurement	What does it measure?	Scalar or vector?	SI unit	Example
distance (d)	the length of a path between two points	scalar	m, km	If you skateboard 10 km [E] of your home, you travelled a *distance* of 10 km.
position ($\vec{d}$)	a specific point relative to a point of origin	vector	m, km	If you skateboard 10 km [E] and return home in a straight line along the same route, your *position* upon returning is 0 km because you are back at your point of origin.

time (t)	when an event occurs	scalar	s, h	You pass a fire hydrant 2 s after you leave your point of origin.
time interval (Δt)	the duration of an event; final time minus the initial time	scalar	s, h	You pass a fire hydrant 2 s after you leave your point of origin. Then, 5 s after you leave your point of origin, you pass a road sign. The *time interval* between these two events is 3 s.
displacement ($\Delta \vec{d}$)	the straight-line distance and direction from one point to another; final position minus the initial position	vector	m, km	At 2 s, you pass the fire hydrant 2 m [E] of your point of origin. At 5 s, you pass the sign at 7 m [E]. Your *displacement* is 5 m [E] during this 3 s time interval.

Why are signs important when using vectors?

Directions are designated as positive or negative when using vectors. North, east, up, and right are positive (+) and south, west, down, and left are negative (–). If a skater travelled from 9 m east of a hydrant to 5 m west of the hydrant, to calculate her displacement, 9 m [E] becomes + 9 m and 5 m [W] becomes –5 m.

$$\Delta \vec{d} = \vec{d}_f - \vec{d}_i$$

$$\Delta \vec{d} = -5 \text{ m} - (+9 \text{ m})$$

$$= -14 \text{ m}$$

Since the negative sign (–) represents west, the skater's displacement is 14 m [W]. ✔

What is uniform motion and how is it represented?

An object in **uniform motion** travels equal displacements in equal time intervals. It does not change speed or direction. A **position-time graph** (like the ones on the next page) shows how an object's position changes over time, allowing its motion to be analyzed. These graphs have the following characteristics:

◆ Time is plotted on the horizontal axis (*x*-axis) and position is plotted on the vertical axis (*y*-axis).

◆ Uniform motion is shown as a straight line.

◆ Real motion is not perfectly uniform. It is useful to use a **best-fit line**, a smooth curve or straight line that most closely fits the general shape outlined by the points, to graph real motion.

✔ Reading Check

2. Why are signs important when using vectors?

_____

_____

_____

_____

Name

Date

Section

8.1

Summary

continued

◆ Positions and times not given as data can be estimated by finding the location corresponding to a specific time and position on the best-fit line. The line can also be extended beyond the first and last points to indicate what might happen beyond the measured data.

What does the slope of a position-time graph tell you?

The **slope** of a graph refers to whether a line is horizontal or goes up or down at an angle. There are three types of slope on a position-time graph:

1. Positive slope: A **positive slope** slants up to the right, indicating that an object's position, from the origin, is increasing with respect to time.

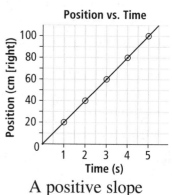
A positive slope

2. Zero slope: **Zero slope** is a straight, horizontal line. It represents an object at rest.

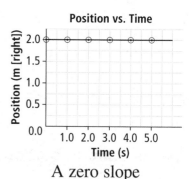
A zero slope

3. Negative slope: A **negative slope** slants down to the right, indicating an object is moving in a negative direction—left, down, west, or south.

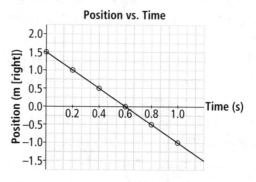
A negative slope

Use with textbook pages 344–347.

Scalars versus vectors

1. Define the following terms.

a) scalar _____

b) vector _____

c) magnitude _____

d) reference point _____

2. Complete the following table.

Quantity	Symbol	SI Unit	Scalar or Vector
time			
time interval			
distance			
position			
displacement			

3. Identify whether the statement is describing a scalar or a vector. Place an "S" for scalar and a "V" for vector in the space provided.

a) _____ A squirrel runs 7 m east of a tree.

b) _____ The school is 5 km from the airport.

c) _____ It took the class 30 minutes to complete the motion lab.

d) _____ A little girl pulls her wagon 10 m west of the tree house.

4. Indicate whether the direction is positive (+) or negative (–).

a) _____

 right

b) _____

west

c) _____

north

d) _____

down

Use with textbook pages 346–349.

Distance, position, and displacement

1. Complete the following table by filling in the blank boxes. In the last column of the table, circle the appropriate word from each pair.

t_i (s)	t_f (s)	Δt (s)	d_i (m)	d_f (m)	Δd_i (m)	Direction of Motion
6.0	7.5		+18.4	+22.6		left/right
	8.5	2.8	+24.3		+5.8	up/down
20.2		18.2		+24.8	−14.3	north/south
12.4	18.8			+46.2	−8.6	east/west

2. Use the following diagram to answer the questions below.

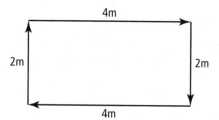

A girl walks 2 m [N], 4 m [E], 2 m [S] and then 4 m [W].

a) What is the total distance the girl travelled? _____

b) What is the displacement of the girl? _____

3. Use the following diagram of a cross-country skier to answer the questions below.

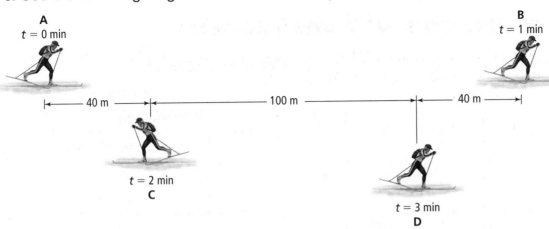

A cross-country skier moves toward the east, then toward the west, and then toward the east again. In other words, the skier moves from position A to B to C to D in 3 minutes.

a) Complete the following tables.

Time	Position
0 min	0 m
1 min	
2 min	40 m [E]
3 min	

Time Interval	Distance Travelled	Displacement
0 min–1 min	180 m	
1 min–2 min		
2 min–3 min		100 m [E]

b) What is the total distance travelled after 3 min? _____

c) What is the skier's displacement at 3 min? _____

Use with textbook page 353–354.

Positive, negative, and zero slopes

Use the following position-time graphs to answer the questions below.

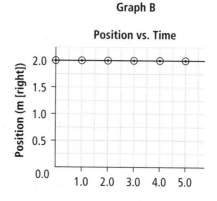

Graph A — Position vs. Time

Graph B — Position vs. Time

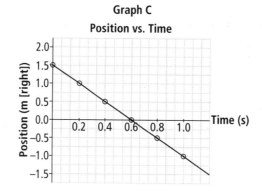

Graph C — Position vs. Time

Match the Description below with the corresponding Graph shown above. Each Graph can be used as often as necessary. Write the correct letter in the space provided.

1. _____ a line with a zero slope

2. _____ a line with a positive slope

3. _____ a line with a negative slope

4. _____ a line that represents uniform motion

5. _____ the motion of an object at rest (not moving)

6. _____ the motion of an object moving to the left of the reference point

7. _____ the motion of an object moving to the right of the reference point

Use with textbook page 350.

Uniform motion

1. Identify each of the situations below as either uniform motion or non-uniform motion.

a) a snowball rolls down a hill _____

b) a man sits on bench watching pigeons _____

c) a woman walks through a crowded mall during the Christmas season

_____

Use the following position-time graph showing the motion of an object, initially moving to the right, to answer the questions below 2 to 4.

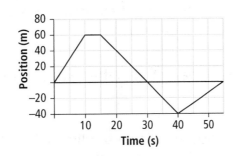

2. For each time interval, describe the slope of the line (positive, negative, or zero) and the motion of the object.

Time Interval	Slope of Line	Description of Motion
0 s–10 s	positive	The object is moving to the right of the origin with uniform motion.
10 s–15 s		
15 s–30 s		
30 s–40 s		
40 s–55 s		

3. During which time interval did the object travel the shortest distance? _____

4. During which time interval did the object travel the longest distance? _____

A student is waiting at a bus stop and starts to pace back and forth. Use the following position-time graph showing the student's motion to answer questions 5 to 11.

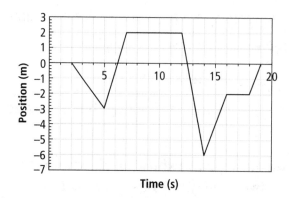

5. During which time intervals is the student standing still?

_____

6. Describe the motion of the student during the time interval 2 s to 5 s.

_____

7. Describe the motion of the student during the time interval 14 s to 16 s.

_____

8. What is the student's position at 7 s? _____

9. What is the student's displacement between 12 s and 14 s? _____

10. What is the total distance covered by the student during the first 16 s? _____

11. What is the student's displacement during the time interval 0 s to 20 s? _____

Use with textbook pages 340–351.

The language of motion

Match the Term on the left with the best Descriptor on the right. Each Descriptor may be used only once.	
Term	**Descriptor**
1. _____ distance 2. _____ position 3. _____ magnitude 4. _____ displacement 5. _____ time interval 6. _____ location 7. _____ Greek letter delta, △	**A.** also known as the origin **B.** the size of a measurement **C.** "change in" or "difference" **D.** a specific point or location relative to a reference point **E.** the total length of a path between two points **F.** the difference between the initial time and the final time **G.** the straight line distance and direction from one point to another

Circle the letter of the best answer.

8. Which of the following units is associated with △t?

   A. s

   B. m

   C. km

   D. m/s

9. Which of the following describes a scalar quantity?

   A. it has direction only

   B. it has magnitude only

   C. it is the size of a quantity

   D. it has both direction and magnitude

Use the following motion diagram representing a ball moving across a horizontal table to answer question 10.

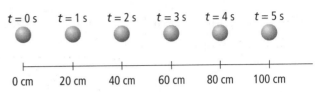

10. Which of the following statements are descriptions of the motion of the ball?

I.	The ball is in uniform motion.
II.	The ball is moving from left to right.
III.	The displacement between t_1 and t_2 is the same as the displacement between t_2 and t_4.

   A. I and II only

   B. I and III only

   C. II and III only

   D. I, II, and III

Use the following position-time graphs to answer question 11.

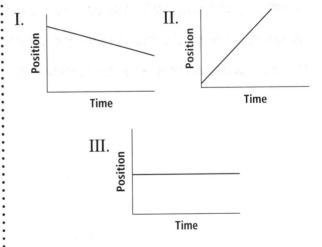

11. Which of the graphs above show uniform motion?

   A. I and II only   **C.** II and III only

   B. I and III only   **D.** I, II, and III

Average Velocity

Textbook pages 362–375

Before You Read

Based on your current knowledge, how do you think speed differs from velocity? Write your answer in the lines below.

_____

_____

State the Main Ideas

Use your own words to express the main ideas covered in this section.

✔ Reading Check

What is velocity?

_____

_____

_____

What is the difference between velocity and speed?

Velocity ($\vec{v}$) is a vector that describes how quickly an object's position changes, as well as the direction of this change. **Speed** (v) is a scalar that measures the magnitude of velocity. Both speed and velocity are measured in metres per second (m/s).

Objects travelling at the same speed can have different velocities. Imagine two escalators travelling at the same speed, one going up, and the other down. Because they are travelling in opposite directions, one of the directions has a negative sign. Thus, they have different velocities. ✔

How is velocity determined on a position-time graph?

Velocity can be determined from the slope of a position-time graph. Where the graph shows a straight line, the velocity is constant. The slope is calculated as follows:

$$\text{Slope} = \frac{\text{rise}}{\text{run}}$$

$$= \frac{\Delta \vec{d}}{\Delta t}$$

The slope shows, on average, how far an object has moved in a certain time interval. In other words, the slope shows the object's average velocity. **Average velocity** ($\vec{v}_{av}$) is the rate of change in position over a time interval. It is almost impossible for an object to move at a perfectly uniform rate. Many factors, such as wind or an uneven surface, may cause the object to slightly speed up or slow down. Average velocity "smoothes out" these changes. It is a vector and includes direction. The slope of a position-time graph can be positive, zero, or negative (see figure on next page).

Name

Date

Section

8.2
Summary

continued

If moving away from the origin is considered positive:
- ◆ a positive slope (a) represents the average velocity of the object moving away from the origin.
- ◆ a horizontal line, which has zero slope (b), represents an object at rest.
- ◆ a negative slope (c) represents the average velocity of the object moving back toward the origin.

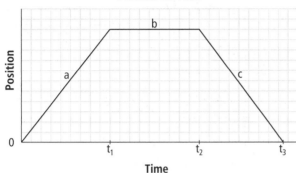

How is average velocity calculated without using a position-time graph?

Since average velocity is the slope of a position-time graph, it can be written as follows:

$$\vec{v}_{av} = \frac{\Delta \vec{d}}{\Delta t} \; ✔$$

By using this relationship, you can calculate the average velocity without analyzing a position-time graph.

Example: A sprinter takes 8.2 s to run forward 75.0 m. What is the sprinter's average velocity?

$$\Delta \vec{d} = +75.0 \text{ m}, \Delta t = 8.2 \text{ s}$$

$$\vec{v}_{av} = \frac{\Delta \vec{d}}{\Delta t}$$

$$= \frac{+75.0 \text{ m}}{8.2 \text{ s}}$$

$$= +9.1 \text{ m/s}$$

Thus, the sprinter ran 9.1 m/s forward.

This equation can also be rearranged to calculate displacement or time.

For displacement:

$$\Delta \vec{d} = (\vec{v}_{av})(\Delta t)$$

For time

$$\Delta t = \frac{\Delta \vec{d}}{\vec{v}_{av}}$$

✔ **Reading Check**

Give the equation for average velocity.

_____

Use with textbook pages 362–366.

Calculating average velocity

1. What is the formula for each of the following quantities?

a) average velocity _____   **b)** displacement _____   **c)** time _____

2. Complete the following table. Use the motion formula to calculate the missing quantities. Show all your work and use the correct units.

Displacement	Time	Average Velocity	Formula Used and Calculation Shown
15.6 m	3 s	5.2 m/s	$\vec{v}_{av} = \dfrac{\Delta\vec{d}}{\Delta t} = \dfrac{15.6}{3} = 5.2$ m/s
357.5 km	6.5 h		
22.6 m		5.65 m/s	
	3.25 h	75 km/h	
12.6 m	3.15 s		
24 km		32 km/h	
	8 s	60 m/s	

3. Complete the following table. Show all your work and use the correct units.

Question	Formula Used and Calculation Shown	Answer
a) A woman wants to paddle 420 m across a lake in her kayak. If she paddles across the lake at an average velocity of 2.8 m/s, how long does it take her to cross?		
b) If a cyclist rides west at 14 m/s, how long would it take her to travel 980 m?		
c) A cheetah runs at a velocity of 30 m/s [E]. If it runs for 8.5 s, what is its displacement?		
d) The Australian dragonfly can fly at 16 m/s. How long does it take to fly 224 m?		
e) The Skyride gondola at Grouse Mountain in North Vancouver takes 8 min to go up the 3 km mountain. What is the average velocity of the gondola?		
f) Due to plate tectonics, the North American and European continents are drifting apart at an average speed of about 3 cm per year. At this speed, how long (in years) will it take for them to drift apart by another 2400 m?		
g) A dragster heading north, reaches a velocity of 628 km/h from rest in 3.72 s. How far did it travel in that time?		

Use with textbook pages 364–367.

Slopes of position-time graphs

1. What does the slope of a line on a position-time graph represent?

_____

2. What does a straight line on a position-time graph represent? _____

3. Define slope. _____

_____

_____

4. What is the formula used to calculate the slope of a straight line?

_____

5. Using the position-time graph, determine the slope of each line segment by completing the following table.

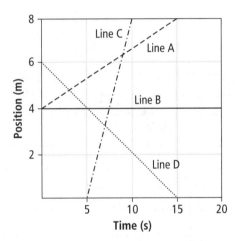

Line	Rise	Run	Slope Calculation	Slope
A				
B				
C				
D				

Use with textbook pages 364–367.

Analyzing position-time graphs

1. Use the following position versus time graph showing a girl's movement up and down the aisle of a store to answer the questions below. The origin is at one end of the aisle.

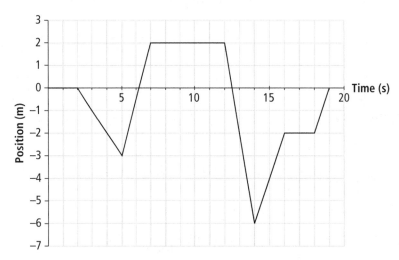

a) Complete the following table. Determine how far the girl travelled during each time interval and calculate the average velocity for each section of the graph.

Time Interval	Displacement	Average Velocity
0 s–2 s		
2 s–5 s		
5 s–7s		
7 s–12 s		
12 s–14 s		
14 s–16 s		
16 s–18 s		
18 s–19 s		
19 s–20 s		

b) When does the girl have a position of –6 m? _____

c) What is the girl's total displacement after 20 seconds? _____

2. Use the following position-time graph, showing the motion of a gymnast on a balance beam, to Match each **Descriptor** below with the corresponding part of the **Graph** shown above. Each part of the Graph may be used as often as necessary. Assume the centre of the balance beam is the reference point (origin).

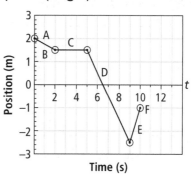

_____ **a)** She stands still for 3 s.

_____ **b)** She moves even faster to the right for 1 s.

_____ **c)** She moves very slowly to the left for 2 s.

_____ **d)** She moves more quickly to the left for 4 s.

_____ **e)** She ends up 1 m left of the centre of the balance beam.

_____ **f)** She starts 2 m to the right of the centre of the balance beam.

3. Use the following position-time graph, showing the motion of two runners, to answer the questions below.

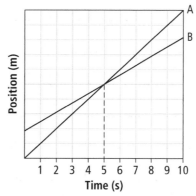

a) What does the *y*-intercept represent? _____

b) Do the runners start at the same place? _____

c) At about 2 s, which runner is running faster? How can you tell?

   _____

d) What occurs at 5 s? _____

e) At 10 s, which runner is ahead? _____

Use with textbook pages 366–367

Constructing and interpreting position-time graphs

1. Use the following data table, showing a car's recorded positions over 7 seconds, to
answer the questions below. Assume 0 m is the reference point.

Time (s)	Position (m)
0	125
1	100
2	75
3	50
4	25
5	0
6	−25
7	−50

a) Label the x-axis with Time (s) and the y-axis with Position (m). Use the grid to plot
the data points from the data table. Draw a best-fit line through the points.

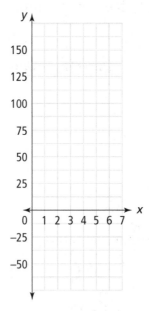

b) When was the car 50 m [E] of the reference point? _____

c) What was the car's position at 1 s? _____

d) Where was the car at 5.5 s? _____

e) What was the car's average velocity between 0 s and 7 s? _____

f) Describe the motion of the car during the time interval 2 s – 4 s.

_____

2. Sketch a position-time graph for each of the following scenarios. If specific time, positions, and velocities are given, label them on the graph. Assume all motion is uniform and in a straight line.

a) A car is travelling north at a velocity of 50 km/h. It slows down to 30 km/h when it enters a school zone.

b) A boy walks away from the kitchen table, 4 m to the right with a velocity of 2 m/s. He spends 6 s getting a bowl of fruit salad out of the refrigerator, and then walks back to the table at a velocity of 1 m/s.

c) At soccer practice, the coach makes the players run back and forth between two lines four times.

Use with textbook pages 362–370.

Average velocity

Use the following position-time graph to answer questions 1 to 4.

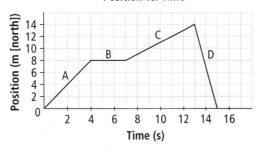

Position vs. Time

Match each Descriptor below with the corresponding part of the Graph shown above.
1. _____ has an average velocity of 0 m/s
2. _____ has an average velocity of 1 m/s [N]
3. _____ has an average velocity of 2 m/s [N]
4. _____ has an average velocity of 7 m/s [S]

5. Which two terms represent a vector quantity and the scalar quantity of the vector's magnitude, respectively?

   A. velocity and speed

   B. time and time interval

   C. acceleration and velocity

   D. position and displacement

6. Which of the following graphs represent the motion of an object with a constant velocity?

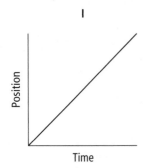

II

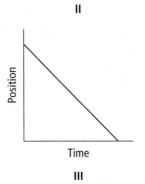

III

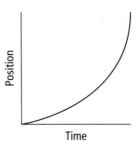

   A. I and II only

   B. I and III only

   C. II and III only

   D. I, II, and III

7. How long would a meteor, with a velocity of 45 km/s, take to travel 120 km through Earth's atmosphere to Earth's surface? Assume no atmospheric friction.

   A. 0.375 s

   B. 2.66 s

   C. 45 s

   D. 5400 s

8. It took 0.45 s for a fastball to reach the batter. If the pitcher is 18 m away from the batter, how fast was the fastball pitch?

   A. 0.025 m/s

   B. 8.1 m/s

   C. 18 m/s

   D. 40 m/s

9. The average velocity of a plane was 600 km/h [N]. How long did it take the plane to travel 120 km?

  A. 0.2 min

  B. 5 min

  C. 12 min

  D. 5.0 h

10. A Canada goose flew 860 km from Washington to British Columbia with an average velocity of 30.5 m/s [N]. Approximately how long did it take the goose to make its journey?

  A. 0.04 h

  B. 7.8 h

  C. 28.2 h

  D. 469.9 h

11. An odometer in a car has a reading of 50 km at the beginning of a trip and a reading of 125 km half an hour later. What is the average speed of the car?

  A. 3.75 km/h

  B. 62.5 km/h

  C. 150 km/h

  D. 250 km/h

Use the following position-time graph of the motion of an object to answer questions 12 to 14.

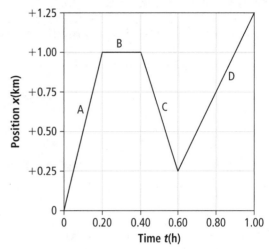

12. During which time interval is the object moving towards the origin?

  A. interval A

  B. interval B

  C. interval C

  D. interval D

13. What is the average velocity between 0.6 h and 1 h?

  A. +0.4 km/h

  B. +1.25 km/h

  C. +1.66 km/h

  D. +2.5 km/h

14. What is the object's displacement between 0.4 h and 0.6 h?

  A. −0.75 km

  B. +0.25 km

  C. +0.75 km

  D. +1.0 km

Describing Acceleration

Textbook pages 380–391

Before You Read

Are you accelerating if you are slowing down? Explain your answer on the lines below.

_____

_____

 Create a Quiz

After you have read this section, create a five-question quiz based on what you have learned. Repeat the quiz until you get all the answers correct.

How can you calculate a change in velocity?

A **change in velocity** ($\Delta \vec{v}$) occurs when the speed of an object changes, or its direction of motion changes, or both. Changes in velocity can be either positive or negative. To find a change in velocity, subtract the initial velocity ($\vec{v}_i$) from the final velocity ($\vec{v}_f$).

$$\Delta \vec{v} = \vec{v}_f - \vec{v}_i$$

How do signs indicate changes in velocity?

North, east, up, and right are considered positive (+) and south, west, down, and left are negative (–). If you slow down from 9 m/s forward (positive) to 2 m/s forward (positive), your change in velocity is as follows:

$$\Delta \vec{v} = \vec{v}_f - \vec{v}_i$$
$$= +2 \text{ m/s} -(+9 \text{ m/s})$$
$$= -7 \text{ m/s}$$

Your change in velocity is 7 m/s opposite the forward motion. Your initial forward direction is *positive*, so your change in velocity is *negative* when you slow down.

What is acceleration?

Acceleration is the rate at which the velocity of a moving object changes. A change in velocity can be a change in either speed or direction. Thus, acceleration occurs when the speed of an object changes, or its direction of motion changes, or both. Acceleration is a rate of change. This means it also takes into account how quickly the velocity changes.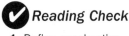

 Reading Check

1. Define acceleration.

_____

_____

_____

Name

Date

Section

9.1

Summary

continued

How do signs indicate changes in acceleration?

Recall that forward motion is defined as positive and backward motion is defined as negative. Different factors can help you decide if an object's acceleration is positive or negative, as shown in the table below: ✔

Factor	Velocity	Acceleration
increase in speed while travelling forward, e.g., accelerating after you have stopped at a stop sign	+ (positive)	+ (positive)
decrease in speed while travelling forward, e.g., applying the brakes on a bicycle	+ (positive)	– (negative)
increase in speed while travelling backward, e.g., a ball falling to earth	– (negative)	– (negative)
no change in speed, e.g., running at a constant speed	constant	0

Note that an object that is slowing down is changing its velocity; therefore, it is accelerating. Acceleration in a direction that is opposite the direction of motion is sometimes called **deceleration.**

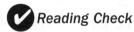

Reading Check

2. What is the acceleration of an object travelling at a constant velocity?

_____

Use with textbook pages 380–386.

Velocity and acceleration

Vocabulary	
constant velocity	positive acceleration
deceleration	same direction
negative	speed
negative acceleration	vector
opposite direction	velocity
positive	

Use the terms in the vocabulary box to fill in the blanks. You may use each term only once.

1. Since velocity is a _____, it is dependent on the _____ of the object and the direction in which the object is moving.

2. A change in velocity is _____ when an object speeds up.

3. A change in velocity is _____ when an object slows down.

4. An object has _____ when it is travelling with uniform motion.

5. Acceleration is the rate of change in _____ .

6. An object has a _____ when its speed is increasing.

7. An object has a _____ when its speed is decreasing.

8. If an object's acceleration is in the _____ as its velocity, the object's speed increases.

9. If an object's acceleration is in the _____ as its velocity, the object's speed decreases.

10. Acceleration that is opposite to the direction of motion is called _____ .

Use with textbook page 382.

Calculating change in velocity

1. Complete the following table by calculating the missing quantities. Positive (+) represents the forward motion. Use the formula $\Delta \vec{v} = \vec{v}_f - \vec{v}_i$. In the last column, describe the change in velocity (e.g. object is slowing down, object is speeding up, or object is in uniform motion).

$\vec{v}_i$	$\vec{v}_f$	$\Delta \vec{v}$	Description of $\Delta \vec{v}$
+ 14 m/s	+ 5 m/s		object is slowing down
+ 8 m/s		0 m/s	
	+ 25 m/s	+ 12 m/s	
+ 20 m/s	− 30 m/s		
− 38 m/s		− 10 m/s	
	− 16 m/s	0 m/s	
− 3 m/s	+ 22 m /s		

2. Use the following data table to calculate the change in velocity for each time interval. Suppose motion toward north is positive (+).

Time (s)	Velocity (m/s)
0	0
10	15
20	28
30	28
40	22
50	12

a)　0 s – 10 s＿＿＿＿＿＿＿＿＿＿＿＿＿＿＿＿＿＿＿＿＿＿＿

b)　10 s – 20 s＿＿＿＿＿＿＿＿＿＿＿＿＿＿＿＿＿＿＿＿＿＿

c)　20 s – 30 s＿＿＿＿＿＿＿＿＿＿＿＿＿＿＿＿＿＿＿＿＿＿

d)　30 s – 40 s＿＿＿＿＿＿＿＿＿＿＿＿＿＿＿＿＿＿＿＿＿＿

e)　40 s – 50 s＿＿＿＿＿＿＿＿＿＿＿＿＿＿＿＿＿＿＿＿＿＿

Use with textbook page 385–386.

Positive, negative, and zero acceleration

1. In each situation described below, identify whether the object or person has positive acceleration, negative acceleration, or zero acceleration.

   a) an airplane taking off _____

   b) a person standing still at a bus stop _____

   c) a bus braking as it approaches a red light

   _____

   d) a person sliding down a water slide with constant velocity

   _____

2. In each illustrated example shown below, identify whether the object or person has positive acceleration, negative acceleration or zero acceleration.

   a) _____   b) _____

   acceleration   acceleration

   velocity   velocity

   c) _____   d) _____

   $\vec{v}_i = +20$ m/s $\vec{v}_f = +6$ m/s $\vec{v}_f = -4$ m/s $\vec{v}_i = -1$ m/s

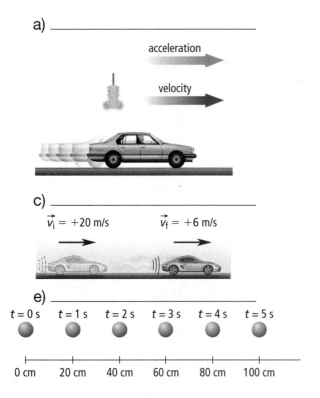

   e) _____   f) _____

   $t = 0$ s $t = 1$ s $t = 2$ s $t = 3$ s $t = 4$ s $t = 5$ s

   0 cm   20 cm   40 cm   60 cm   80 cm   100 cm

Use with textbook pages 382–386.

Describing acceleration

Match the Term on the left with the best Descriptor on the right. Each Descriptor may be used only once.	
Term	**Descriptor**
1. _____ acceleration 2. _____ deceleration 3. _____ constant velocity 4. _____ change in velocity	**A.** rate at which an object changes its velocity **B.** an object travelling with uniform motion in a straight line **C.** acceleration in a direction that is opposite to the direction of motion **D.** change that occurs when the speed of an object changes, or its direction of motion changes, or both

Circle the letter of the best answer.

5. The tortoise moved slowly and steadily. The hare ran quickly, then fell asleep near the finish line. Identify the following quantities as scalar or vector, respectively: the average velocity of the tortoise during the entire race, the acceleration of the hare during the first 2 minutes of the race and the time it takes for the tortoise and the hare to finish the race.

   A. vector, vector, scalar

   B. scalar, scalar, vector

   C. vector, scalar, vector

   D. scalar, vector, scalar

6. If the acceleration of an object is in the opposite direction as the velocity, which of the following happens?

   A. the object speeds up

   B. the object slows down

   C. the object remains at rest

   D. nothing happens to the object

7. Suppose an object moving forward changes its velocity from 15 m/s to 6 m/s. What is the change in velocity?

   A. –21 m/s

   B. –9 m/s

   C. 9 m/s

   D. 21 m/s

8. Which of the following have an acceleration in the same direction as the object's motion?

I.	$\vec{v}_i$ = 3.5 m/s [east]; $\vec{v}_f$ = 7.5 m/s [east]
II.	$\vec{v}_i$ = 45 km/h [north]; $\vec{v}_f$ = 60 km/h [north]
III.	$\vec{v}_i$ = 15 m/s [right]; $\vec{v}_f$ = 10 m/s [left]

   A. I and II only

   B. I and III only

   C. II and III only

   D. I, II, and III

9. Which of the following are true about the acceleration of an object travelling in a straight line?

I.	When an object's speed is constant, the object has zero acceleration.
II.	When an object's speed is increasing, the object has a positive acceleration.
III.	When an object's speed is decreasing, the object has a negative acceleration.

   A. I and II only

   B. I and III only

   C. II and III only

   D. I, II, and III

Calculating Acceleration

Textbook pages 392–405

Before You Read

How do you think a velocity-time graph might differ from the position-time graph you learned about in the previous chapter? Write your answer on the lines below.

_____

_____

Draw a Graph

Draw a velocity-time graph for an object experiencing positive, zero, and negative acceleration.

✔ *Reading Check*

1. Write the equation for acceleration.

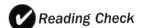

How is acceleration determined on a velocity-time graph?

A **velocity-time graph** represents the motion of an object with changing velocity. The slope of a velocity-time graph gives the object's acceleration, which is measured in m/s². When a best-fit line passes through all data points, the object's velocity is changing at a constant rate and it experiences **constant acceleration**. However, since not all the velocities may be directly on the best-fit line, the slope is referred to as **average acceleration** ($\vec{a}$).

If north is considered positive, for lines above the *x*-axis:

♦ a positive slope (a) represents the average acceleration of an object that increases speed at a constant rate while travelling north. Acceleration is constant and positive.

♦ zero slope (b) represents an object travelling north at a constant speed. It is not accelerating.

♦ a negative slope (c) represents an object that decreases speed at a constant rate while travelling north. Acceleration is constant and negative. Velocity is positive.

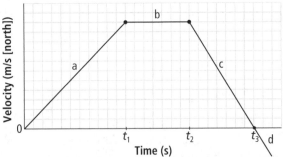

Velocity vs. Time

A line below the *x*-axis (d) represents increasing speed at a constant rate toward the south. Acceleration is constant and negative. Velocity is negative.

How is acceleration calculated without using a velocity-time graph?

Recall that average acceleration is the slope of a velocity-time graph:

$$\text{Slope} = \frac{\text{rise}}{\text{run}}$$
$$= \frac{\Delta \vec{v}}{\Delta t}$$

This textbook only considers situations where acceleration is constant. This means average acceleration is actually the same as acceleration at any instant.

$$\vec{a} = \frac{\Delta \vec{v}}{\Delta t} \ ✔$$

This equation can be rearranged to calculate velocity or time.

For velocity:

$$\Delta \vec{v} = \vec{a} \Delta t$$

For time:

$$\Delta t = \frac{\Delta \vec{v}}{\vec{a}}$$

What is the relationship between gravity and acceleration?

When an object falls near Earth's surface, the force of **gravity** pulls it downward. Consider a ball being thrown straight up into the air, where "up" is positive.

◆ On the way up, the ball's velocity is decreasing. The ball is slowing down, so its acceleration is negative.

◆ At its maximum height, the ball's velocity is zero for an instant since the direction of the ball is changing. (Because the ball's velocity is still changing, the ball is accelerating although its velocity is zero for an instant.)

◆ When the ball starts to come down, its speed increases. However, its velocity is negative because the ball is heading "down." The ball's acceleration is negative.

How does air resistance influence acceleration due to gravity?

Objects fall at different rates because of **air resistance**, a friction-like force. In the absence of air resistance, all objects, regardless of their weight, fall with the same constant acceleration of 9.8 m/s² downward. This is **acceleration due to gravity** (g). In many situations, the air resistance acting on a falling object is so small that we can assume the object has a constant acceleration of –9.8 m/s², where up is positive. ✔

✔ **Reading Check**

2. What is the value of acceleration due to gravity on Earth?

_____

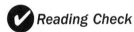

Name _______________ Date _______________

Applying Knowledge
Section 9.2

Use with textbook pages 396–400.

Calculating acceleration

1. What is the formula for each of the following quantities?

(a) acceleration ________ (b) change in velocity ________ (c) time interval ________

2. Complete the following table. Use the motion formula to calculate the missing quantities. Show all your work and use the correct units.

Change in Velocity	Time	Acceleration	Formula Used and Calculation Shown
140 m/s	8 s	17.5 m/s²	$\vec{a} = \frac{\Delta \vec{v}}{\Delta t} = \frac{140}{8} = 17.5$ m/s
–60 km/h	4 h		
120 km/h		48 km/h²	
	15 s	–3.5 m/s²	
12 m/s	2.5 s		
–25 m/s		–12.5 m/s²	
	9.6 h	5 km/h²	

3. Solve each problem using the appropriate motion formula. Show all your work and use the correct units.

a) A car moving north goes from 5.56 m/s to 63.9 m/s in 7.5 s. What is the acceleration?

____

b) If a sprinter starts a race and has an acceleration of 2.4 m/s² in 2.5 s, what is his final velocity, assuming the initial velocity is 0 m/s?

____

c) A rock accelerates at –9.8 m/s² when falling. How long does it take to change its velocity from –4.5 m/s to –19.4 m/s?

____

d) A satellite released from a stationary space shuttle accelerates to +68 m/s² in 25 s. What is its change in velocity?

____

Use with textbook pages 394–396.

Analyzing velocity-time graphs

1. What is the meaning of each of the following features of a velocity-time graph?

(a) the slope of the line _____

(b) a line above the *x*-axis _____

(c) a line below the *x*-axis _____

(d) a line with a positive slope _____

(e) a line with a negative slope _____

(f) a horizontal section of the graph _____

(g) a point where the line crosses the *x*-axis _____

Use the following velocity-time graph representing the motion of a ball moving to the right on a table to answer questions 2 and 3.

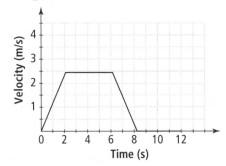

2. Complete the following table. Describe the slope, acceleration, and velocity of the ball (e.g. positive, negative, or zero).

MOTION OF A BALL			
Time Interval	Slope	Acceleration	Velocity
0 s – 2 s			
2 s – 6 s			
6 s – 8 s			
8 s – 12 s			

3. Describe the motion of the ball at each time interval.

(a) 0 s – 2 s _____

(b) 2 s – 6 s _____

(c) 6 s – 8 s _____

(d) 8 s – 12 s _____

Use with textbook pages 395–396.

Sketching and interpreting velocity-time graphs

1. Complete the following table. What is the slope (e.g. positive, negative, or zero) of each velocity-time graph? State whether the graph shows positive acceleration, negative acceleration, or zero acceleration.

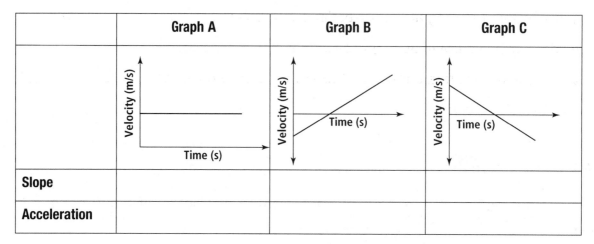

	Graph A	Graph B	Graph C
Slope			
Acceleration			

2. Sketch a velocity-time graph for each scenario given below.

	Positive Acceleration	**Negative Acceleration**
Positive Velocity		
Negative Velocity		

3. (a) Sketch a velocity-time graph of a field trip to the science museum, showing all the stages (i to v) listed below.

   i. the bus is stationary (has an initial velocity of zero) as the students board the bus at school

   ii. the bus has constant acceleration as it leaves the school

   iii. the bus is travelling at the speed limit with uniform motion on the highway

   iv. the bus slows down as it approaches some traffic

   v. the bus comes to a complete stop at the science museum

(b) Identify the sections of the velocity-time graph with positive, negative, and zero slope.

i: _____

ii: _____

iii: _____

iv: _____

v: _____

(c) Identify the stages of the field trip with positive, negative, and zero acceleration.

i: _____

ii: _____

iii: _____

iv: _____

v: _____

Use with textbook pages 392–400.

Calculating acceleration

Match the Descriptor on the left with the best Velocity-Time Graph on the right. The Velocity-Time Graphs represent the motion of a car heading north. Each Velocity-Time Graph may be used only once.	
Descriptor	**Velocity-Time Graph**
1. _____ The car is stopped. **2.** _____ The car is accelerating. **3.** _____ The car is slowing down. **4.** _____ The car is travelling at constant velocity.	**A.** **B.** **C.** **D.**

A. Velocity (m/s) / Time (s)

B. Velocity (m/s) / Time (s)

C. Velocity (m/s) / Time (s)

D. Velocity (m/s) / Time (s)

5. Acceleration is represented by the slope of a

A. velocity-time graph

B. position-time graph

C. distance-time graph

D. acceleration-time graph

6. A meteor goes from +1.0 km/s to +2.2 km/s in 0.04 s. What is its acceleration?

A. 0.03 km/s²

B. 30 km/s²

C. 55 km/s²

D. 80 km/s²

Use the following velocity-time graph to answer question 7.

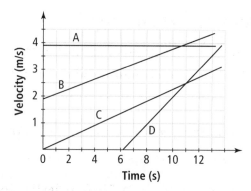

7. Which line represents an object with the greatest acceleration?

A. A

B. B

C. C

D. D

Use the following velocity-time graph to answer question 8.

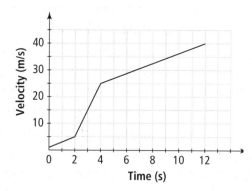

8. What is the object's acceleration between the time interval 2 s and 4 s?

A. +1 m/s²

B. +10 m/s²

C. +20 m/s²

D. +40 m/s²

Temperature, Thermal Energy, and Heat

Textbook pages 424–435

Before You Read

We often use the terms heat and temperature interchangeably. Do you think they mean the same thing? Explain your reasoning in the lines below.

_____

_____

How is energy associated with moving particles?

The **kinetic molecular theory** explains that particles in matter are in constant motion. **Kinetic energy** is the energy of a particle or an object due to its motion. When particles collide, kinetic energy is transferred between them. The particles of a substance move at different speeds depending on the state of the substance. The particles of a gas have more kinetic energy than those of a liquid and move more quickly. The particles of a liquid have more kinetic energy than those of a solid.

Kinetic energy is not the only energy associated with moving particles. **Potential energy** is stored energy that has the *potential* to be transformed into another form of energy, such as kinetic energy. A good example is the gravitational attraction between Earth and the textbook you are holding. As you lift the textbook, its gravitational potential energy increases. The book has a greater distance to fall, so more energy will be transformed into kinetic energy if it does. On the other hand, the lower you hold the book, the less gravitational potential energy it has. At a lower height, less energy will be transformed into kinetic energy if the book falls. Similarly, there are attractive electrical forces between atoms and molecules. The pull of these attractive forces also gives particles potential energy. ✔

How is kinetic energy measured?

Kinetic energy is measured in terms of temperature, thermal energy, and heat.
1. Temperature is a measure of the *average kinetic energy* of all the particles in a sample of matter. As the particles'

Mark the Text

Check for Understanding

As you read this section, be sure to reread any parts you do not understand. Highlight any sentences that help make concepts clearer for you.

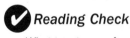
Reading Check

What two types of energy are associated with moving particles?

_____

_____

Name

Date

Section

10.1

Summary

continued

average kinetic energy increases, the temperature of the sample also increases, and vice versa. For example, particles in a glass of cold water move more slowly than, and therefore have less kinetic energy than, particles in a cup of hot water.

Three different scales are used to measure temperature: Fahrenheit, Celsius, and Kelvin.

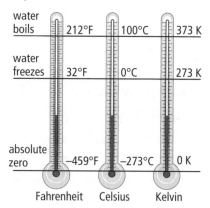

2. Thermal energy is the *total energy* of all the particles in a solid, liquid, or gas. A hot bowl of soup has more thermal energy when it is first served than after it cools. So far this is similar to temperature. However, since thermal energy includes the energy of all of the particles in a sample of matter, a large bowl of soup has more thermal energy than a small bowl of soup at the same temperature. In fact, a swimming pool of lukewarm water has more thermal energy than a small cup of hot tea.

3. Heat is the amount of *thermal energy* that transfers from an area or object of higher temperature to an area or object of lower temperature. Heat can be transferred in three ways:

1. Conduction: **Conduction** describes heat transfer that occurs when faster moving particles collide with slower moving particles. During conduction, heat is transferred from matter with a higher temperature and greater kinetic energy to matter with a lower temperature and less kinetic energy. For example, if a metal spoon that is at room temperature is placed in a pot of boiling water, heat will be transferred to the spoon by conduction and it will become hot. Materials often conduct heat at different rates. Metals, for example, are good thermal conductors, while wood and air are not.

2. Convection: **Convection** is the transfer of heat within a fluid, where the fluid actually moves from one place to another. Unlike conduction, convection transfers matter as well as heat. A boiling pot of water provides a good example of how convection works. As the water at the bottom of the pot heats up, the molecules begin to move faster and their kinetic energy increases, causing them to spread apart. The water expands and becomes less dense than the surrounding water. As a result, it rises to the surface, where it cools, contracts, and sinks— only to be reheated and circulated again. This movement of a fluid due to differences in density is called a **convection current**.

3. Radiation: **Radiation** is the transfer of heat by electromagnetic waves that carry radiant energy. One type of radiation associated with heat transfer is called **infrared radiation**, or heat radiation. This is the heat transfer you experience when you stand close to a campfire. The campfire is emitting electromagnetic waves toward your body, causing you to feel warmth. Similarly, everything around you experiences heat transfer as a result of **solar radiation** from the Sun, which includes many different types of electromagnetic waves. ✅

What are Earth's energy sources?

Earth receives energy from three main sources:
1. Solar radiation, including visible light, infrared radiation, and other types of radiation, comes from the Sun.

2. Residual thermal energy from when Earth was formed is slowly released.

3. Decay of underground radioactive elements produces energy.

✔ *Reading Check*

What is the difference between convection and radiation?

_____

_____

_____

_____

Use with textbook pages 424–431.

Kinetic molecular theory and temperature

1. Define the term kinetic energy.

_____

2. Complete the following table by describing the three states of matter in terms of the space between the particles, speed of movement of the particles, and relative amount of kinetic energy.

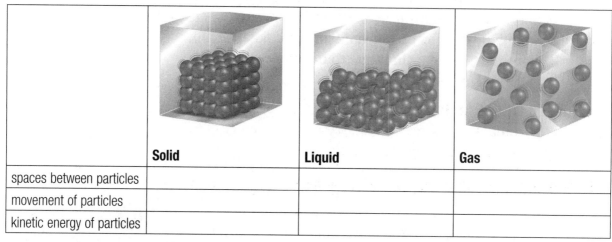

	Solid	Liquid	Gas
spaces between particles			
movement of particles			
kinetic energy of particles			

3. Define the term temperature.

_____

4. In the diagrams below, draw arrows to show how fast the water molecules are moving and in what direction they move in hot and cold water.

hot water

cold water

5. Three different scales are used to measure temperature. Complete the table below comparing the measurements for absolute zero, freezing of water, and boiling of water on the various scales.

	Fahrenheit	Celsius	Kelvin
absolute zero			
water freezes			
water boils			

Name Date

Comprehension
Section 10.1

Use with textbook pages 426–431.

Thermal energy, kinetic energy, potential energy

1. What is thermal energy?

____

2. What is kinetic energy?

____

3. What is potential energy?

____

4. What happens to the thermal energy of an object as its temperature rises?

____

5. What happens to molecules as their kinetic energy increase?

____

6. What happens to molecules as their potential energy increases?

____

7. What is heat?

____

8. Give an example that illustrates the above definition of heat.

____

____

____

9. State three ways in which thermal energy is transferred.

____

Name _______________ Date _______________

Applying Knowledge
Section 10.1

Use with textbook pages 427–431.

Thermal energy transfer

1. Using the illustrations, complete the following table.

	Type of thermal energy transfer	What is happening in the diagram
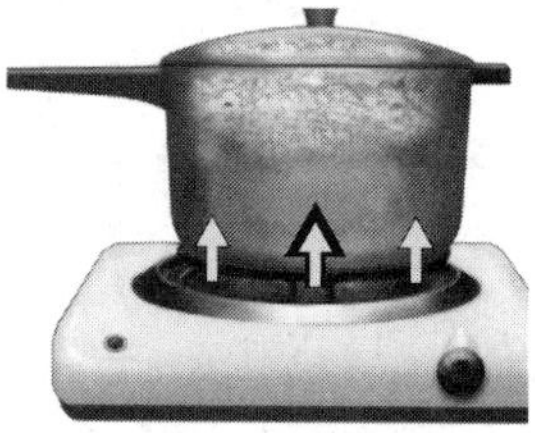		
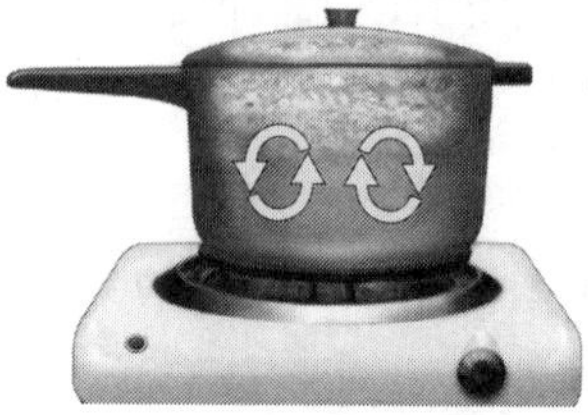		
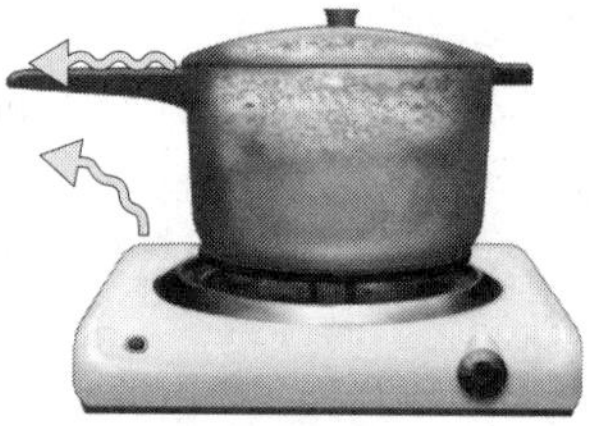		

2. What materials are good thermal conductors?

3. Give three examples of materials that are considered to be insulators.

4. Explain what causes the movement of the liquid in a lava lamp.

5. What is radiant energy?

Temperature, thermal energy, and heat

Use with textbook pages 424–431.

Match each Term on the left with the best Descriptor on the right. Each Descriptor may be used only once.	
Term	**Descriptor**
1. conduction 2. convection 3. electromagnetic spectrum 4. heat 5. kinetic energy 6. kinetic molecule theory 7. temperature 8. thermal energy	**A.** the transfer of thermal energy within a fluid and with the movement of fluid from one place to another **B.** the theory that all matter is composed of particles moving constantly in random directions **C.** the transfer of energy by waves travelling outward in all directions from a source **D.** the transfer of thermal energy from one substance to another or within a solid by direct contact of particles **E.** the total energy of all the particles in a solid, liquid, or gas **F.** a measure of the average kinetic energy of all the particles in a sample of matter **G.** the transfer of thermal energy from an area or object of high temperature to an area or object of low temperature **H.** the energy of a particle or object due to its motion.

Circle the letter of the best answer.

9. As the temperature of an object decreases, the kinetic energy of the object

   A. decreases

   B. increases

   C. remains the same

   D. fluctuates

10. Which of the following best describes heat?

    A. stored energy of an object

    B. transfer of thermal energy

    C. energy of a particle due to its motion

    D. total energy of all particles involved

11. A temperature reading of 273° Kelvin is equivalent to

    A. 0°C

    B. 100°C

    C. 212°F

    D. – 459°F

12. Which type of thermal energy accounts for the movement of clouds?

    A. heat

    B. conduction

    C. convection

    D. radiation

13. Which of the following are sources of thermal energy?

I.	Earth's formation
II.	radioactive decay
III.	humans

    A. I only

    B. II only

    C. I and II only

    D. I, II, and III

Section
10.2
Summary

Energy Transfer in the Atmosphere

Textbook pages 436–459

Before You Read

What do you think causes wind? Write your thoughts in the lines below.

_____

_____

 Mark the Text

In Your Own Words

Highlight the main idea in each paragraph. Stop after each paragraph and put what you have read into your own words.

 **Reading Check**

Name the five layers of the atmosphere.

_____

_____

_____

_____

_____

What is Earth's atmosphere like?

Many planets have **atmospheres**, layers of gases that extend above a planet's surface. Earth's atmosphere is made up of five layers: from lowest to highest, they are the troposphere, stratosphere, mesosphere, thermosphere, and exosphere. These layers differ in chemical composition, average temperature, and density. ✔

The troposphere is the layer nearest the surface of the Earth. Almost all water vapour and dust in the atmosphere is found here. The average temperature near Earth is 15°C but at the top of the troposphere is –55°C. 99% of the gases in the troposphere are nitrogen and oxygen.

The stratosphere has dry air and an average temperature of about –55 °C at the bottom and 0 °C at the top. The ozone layer, which absorbs much of the ultraviolet radiation from the Sun, is in the stratosphere.

Temperatures in the mesosphere can reach as low as –100 °C. Every day, small pieces of dust and meteors rush through the mesosphere.

Temperatures in the thermosphere can reach 1500 °C to 3000 °C. The northern lights, or aurora borealis, are a result of charged particles colliding in the thermosphere.

The boundaries of the exosphere are not well defined, and this layer merges with outer space.

The atmosphere is constantly changing, due to many factors, including the Sun's rotation and the effects of day and night.

How is the atmosphere warmed?

Solar radiation transfers heat to Earth. The amount of solar radiation that reaches a certain area is called **insolation**.

Higher latitudes receive less insolation due to a greater **angle of incidence**. The angle of incidence is the angle that occurs between a ray reaching a surface and a line perpendicular to that surface. It increases with latitude.

Angle of incidence

Very little solar radiation heats the atmosphere directly. Solar radiation arrives in short wavelengths, some of which pass through the atmosphere to Earth's surface, where they are absorbed. Earth's surface reradiates some of this energy as longer, infrared waves. The atmosphere absorbs this infrared radiation and convection transfers the thermal energy throughout the atmosphere.

Earth has a **radiation budget** that keeps incoming and outgoing energy in balance. Incoming short-wave solar radiation is reflected and absorbed to various degrees. **Albedo** describes the amount of radiation reflected by a surface. Forested regions and other dark areas (low albedo), for example, will absorb more radiation than areas covered in ice and snow (high albedo).

What is atmospheric pressure?

Atmospheric pressure is the pressure exerted by the mass of air above any point on Earth's surface. Atmospheric pressure is measured with a **barometer** in **Kilopascals (kPa)**. As the atmospheric pressure changes, a capsule of flexible metal in an aneroid barometer expands or contracts. Kilopascals measure the force per one square metre. Changes in atmospheric pressure occur as a result of the following:

1. Altitude: As altitude increases, atmospheric pressure decreases.

2. Temperature: Warm air is less dense than cold air, resulting in lower atmospheric pressure.

3. Humidity: **Humidity** is a measurement that describes the amount of water vapour in air. The greater the humidity, the lower the atmospheric pressure.

High pressure system	Low pressure system
air cools and becomes more dense	air warms and becomes less dense
air mass contracts, draws in surrounding air, and sinks	air mass expands and rises
due to weight of extra air, atmospheric pressure increases; high pressure air moves outward toward areas of low pressure, creating wind	air pressure at Earth's surface decreases and draws in air from areas of high pressure, creating wind
wind flows clockwise in the northern hemisphere	wind flows counterclockwise in the northern hemisphere
air becomes warmer and drier as it sinks, bringing clear skies	water vapour condenses as air cools, bringing wet weather

What is an air mass?

An **air mass** is a parcel of air with similar temperature and humidity throughout. Conditions in an air mass become like Earth's surface below it. When an air mass cools over a cold region, a high pressure system forms. Air masses that travel over warm regions may develop into low pressure systems. The boundary between two air masses is called a **front**. An approaching front means a change in the weather. The extent of the change depends on the amount of difference between conditions in the two air masses.

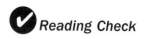 **Reading Check**

What is weather?

_____

_____

_____

_____

What is weather?

Weather is the condition of the atmosphere in a specific place and at a specific time. Weather describes all aspects of the atmosphere, including temperature, atmospheric pressure, humidity, and wind speed and direction. Weather is closely connected to heat transfer in the atmosphere. As heat is transferred, convection moves air and thermal energy throughout the troposphere, causing various kinds of weather.

Several types of extreme weather occur on Earth, including thunderstorms, tornados, and tropical cyclones.

A tornado is a violent funnel-shaped column of air. It is found when high altitude winds meet large thunderstorms. Surface winds caused by tornadoes can reach 400 km/h. Tropical cyclones, or hurricanes, result from the exchange of thermal energy in the tropics. Warm moist air is lifted high into the atmosphere. As rain is produced, thermal energy is released. Warm air rushes to replace the rising air, and the Coriolis effect forces the air to rotate. The result is a massive, spinning storm.

How is wind generated on Earth?

Wind is the movement of air from an area of higher pressure to an area of lower pressure. Geographic features such as mountains, oceans, and lakes greatly affect the characteristics of **local winds**. **Prevailing winds** are winds that are typical for a certain region. Over long distances, wind is also affected by Earth's rotation. The **Coriolis effect** is a change in the direction of moving air, water, or objects due to Earth's rotation. The Coriolis effect and convection currents (rising warm air and sinking cool air) result in three major global wind systems: the trade winds, the prevailing westerlies, and the polar easterlies.

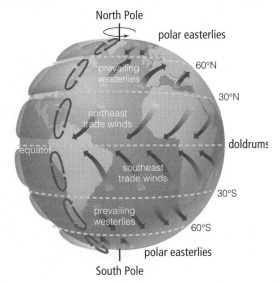

Jet streams form in the upper troposphere due to convection currents and become bands of fast-moving air in the stratosphere. They are so strong that airline pilots try to fly with them.

Use with textbook pages 436–440.

The Earth's atmosphere

Answer the questions below.

1. What is "air"?

_____

2. What two gases make up 99 percent of dry air?

_____

3. What factors cause the atmosphere to constantly change?

_____

_____

4. The Earth's atmosphere is made up of five layers. Each layer differs in average altitude, chemical composition, average temperature, and density.

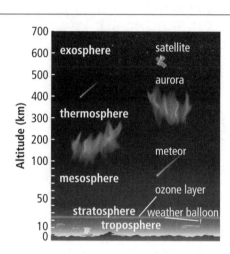

Complete the table below summarizing the characteristics of each layer of the Earth's atmosphere.

Layer	Altitude above sea level	Average temperature	Factors affecting composition
troposphere			
stratosphere			
mesosphere			
thermosphere			
exosphere			

Use with textbook pages 443–448.

What is weather?

1. Define the term weather.

_____

2. What moves air and thermal energy throughout the troposphere?

_____

3. Explain how an aneroid barometer works.

_____

_____

4. What is the SI unit for atmospheric pressure? What does it represent?

_____

5. What causes the sensation of your ears "popping" when you are flying in an airplane?

_____

6. Explain how the following factors affect atmospheric pressure.

(a) temperature increases

_____

(b) warm air mass pushes into region of cold air

_____

(c) cold air mass pushes into a region of warm air

_____

7. Compare the terms wind and air mass.

_____

_____

8. What weather pattern occurs when a high pressure system forms?

_____

9. What weather pattern occurs when a low pressure system forms?

_____

Use with textbook pages 448–454.

Weather patterns

Prevailing Winds of British Columbia

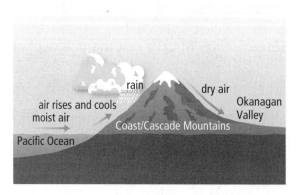

1. Observe the above figure, which depicts the prevailing winds on the west coast. With prevailing winds occurring,

   (a) what would the weather pattern be in Vancouver, British Columbia?

   _____

   (b) what would the weather pattern be in Calgary, Alberta?

   _____

2. Using the diagram, illustrate the Coriolis effect by adding arrows to depict the directions that winds travel in the northern hemisphere and southern hemisphere.

   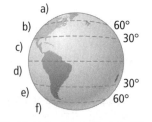

   Wind Patterns Due to Coriolis Effect

3. Use the vocabulary terms that follow to label Earth's global wind systems on the diagram below. Some terms may be used more than once.

   ◆ northeast trade winds
   ◆ polar easterlies
   ◆ southeast trade winds
   ◆ prevailing westerlies

   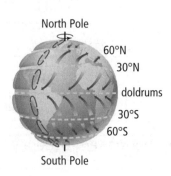

4. What types of weather will occur in the figures below?

(a)

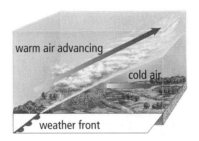

(b)

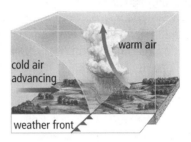

_____   _____

5. Draw a series of diagrams to illustrate the developmental stages of a tornado.

(a) funnel cloud develops	(b) funnel cloud becomes vertical and touches the ground	(c) tornado moves along ground

6. Describe what causes a hurricane to develop.

_____

_____

Energy transfer in the atmosphere

Match each Term on the left with the best Descriptor on the right. Each Descriptor may be used only once.	
Terms	**Descriptor**
1. _____ atmospheres **2.** _____ Coriolis effect **3.** _____ hurricanes **4.** _____ kilopascals (kPa) **5.** _____ ozone layer **6.** _____ prevailing winds **7.** _____ sea breezes **8.** _____ tornado	**A.** local winds caused by the different rates at which land and water transfer thermal energy **B.** tropical cyclones **C.** layers of gases that extend beyond a planet's surface **D.** a change in the direction of moving air, water, or objects due to Earth's rotation **E.** violent, funnel-shaped column of rotating air that touches the ground **F.** winds that are typical for a certain region **G.** the SI unit that measures the vertical force of atmospheric pressure per unit area **H.** the atmospheric layer that absorbs much of the ultraviolet radiation from the Sun

Circle the letter of the best answer.

9. In which layer of the Earth's atmosphere is the ozone layer found?

   A. exosphere

   B. mesosphere

   C. stratosphere

   D. troposphere

10. Where are jet streams found?

    A. exosphere

    B. mesosphere

    C. stratosphere

    D. troposphere

11. What causes an offshore breeze to develop?

    A. land mass has warmed up

    B. land mass has cooled down

    C. ocean mass has warmed up

    D. ocean mass has cooled down

12. A high pressure weather system would result in what type of weather developing over a geographic area?

    A. rainy weather

    B. tornado

    C. clear skies

    D. windy conditions

Natural Causes of Climate Change

Textbook pages 464–481

Before You Read

What is the difference between climate and weather? Write your ideas in the lines below.

_____

_____

What is climate?

Climate describes the average conditions of the atmosphere in a large region over 30 years or more. Climate includes such characteristics as clouds and precipitation, average temperature, humidity, atmospheric pressure, solar radiation, and wind. Climate can refer to conditions in a region as small as an island or to conditions across an entire planet. Because of its varied geography, British Columbia has a range of climates. A **biogeoclimatic zone** is a region with a certain type of plant life, soil, geography, and climate. British Columbia has 14 biogeoclimatic zones.

How do scientists determine past and current climatic change?

Geologic evidence shows that, throughout its history, Earth has undergone many climatic changes, including ice ages and periods of warming. **Paleoclimatologists** study fossils and sediments or gather information about glaciers to help them understand climatic change. They examine **ice cores** to determine what types and amounts of gases existed in the atmosphere when the ice was formed. Ice core data have been used to estimate the concentration of carbon dioxide gas that was in the atmosphere over the past 650 000 years, allowing scientists to estimate past climatic conditions. Scientists draw conclusions about current climatic changes by observing current climate and by comparing their observations with evidence of past climates.

 **Mark the Text**

Identify Concepts

Highlight each question head in this section. Then use a different colour to highlight the answers to these questions.

Which factors affect climate?

The processes that contribute to climate change are complex and include factors that affect Earth's radiation budget and heat transfer around the globe. Several factors affect climate:

1. The composition of Earth's atmosphere: **Greenhouse gases** in the atmosphere absorb and emit radiation as thermal energy, increasing Earth's temperature. The more greenhouse gases, the higher the temperature of our atmosphere. ✔

2. Earth's tilt, axis of rotation, and orbit around the Sun: Earth experiences seasons due to the combination of its tilt and orbit. Seasonal changes are most extreme when Earth's tilt is greatest (the angle of Earth's tilt varies between 22.1° and 24.5° in cycles of about 41 000 years). Changes in Earth's axis of rotation also affect the angle of incidence of the Sun's rays. Variation in the shape of Earth's orbit changes its distance from the Sun and the amount of solar radiation that reaches Earth's surface. In addition, Earth's rotation also has a wobble, which will affect the angle of incidence of the Sun's radiation over a period of thousands of years.

3. The water cycle: The **water cycle** describes the circulation of water on, above, and below Earth's surface. High temperatures increase the evaporation of water (the most abundant greenhouse gas) and the capacity of air to hold water vapour. As surface temperatures rise, so does the amount of water vapour in the atmosphere. As the atmosphere holds more water vapour, it traps more thermal energy, resulting in a further increase in temperature. As temperatures continue to rise, glaciers and ice shelves melt, causing sea levels to rise around the world.

4. Ocean currents: The sinking and rising of deep ocean waters produces convection currents that act as a global conveyer belt that transports water—and thermal energy—around Earth. Surface currents, caused in part by the Coriolis effect, exchange heat with the atmosphere, so these currents also influence both weather and climate.

✔ **Reading Check**

How do greenhouse gases increase Earth's temperature?

_____

_____

_____

_____

Natural Causes of Climate Change

Periodically, surface waters off the coast of Ecuador and Peru get unusually warm, a phenomenon known as an **El Niño** event. Unusually weak westerly trade winds allow warm water in the western Pacific to move eastward. This prevents cold water from upwelling, and triggers changes in weather across much of North America. In contrast, in a **La Niña** event, stronger than normal westerly winds allow cooler-than-normal waters to come to the surface in the eastern Pacific Ocean. This brings cooler temperatures to northwestern North America. Both El Niño and La Niña affect climate in North America. The variation in the winds, including El Niño and La Niña events, is known as El Niño-Southern Oscillation.

5. The carbon cycle: The carbon cycle maintains the balance of carbon dioxide in the atmosphere. Carbon dioxide is an important greenhouse gas. **Carbon sinks**, such as the deep ocean, shelled organisms, and forests, remove carbon dioxide from the atmosphere. Carbon in ocean waters is converted to carbonates, an important ingredient in the shells of many marine organisms. **Carbon sources**, such as weathering and decaying vegetation, add carbon dioxide to the atmosphere. ✔

6. **Catastrophic events**: Large-scale disasters, such as volcanic eruptions and meteor impacts, add dust, debris, and gases high into the atmosphere. They reflect and absorb solar radiation, causing the atmosphere below to cool.

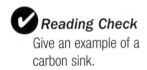

Reading Check

Give an example of a carbon sink.

_____

Use with textbook pages 464–475.

Natural causes of climate change

Vocabulary	
carbon sink	natural greenhouse effect
catastrophic events	paleoclimatologists
climate	shape
convection currents	tilt
Coriolis effect	water vapour
El Niño-Southern Oscillation	weathering
	wobble

Use the terms in the vocabulary box to fill in the blanks. Use each term only once.

1. _____ describes the average conditions of the atmosphere in a large region over 30 years or more.

2. _____ gather information about glaciers using ice cores to determine what types and amounts of gases existed in the atmosphere when the ice was formed.

3. Life on Earth is adapted to the conditions provided by the _____, which balances incoming solar radiation and outgoing heat.

4. The combined effects of _____, _____, and the _____ of Earth's orbit can be linked to the cooling of the global climate in the past and the cause of the ice ages.

5. _____ is the most abundant greenhouse gas in the atmosphere.

6. _____ in the oceans transport large amounts of heat around the globe.

7. Currents of air or water are deflected to the right in the northern hemisphere and to the left in the southern hemisphere due to the _____.

8. The variation in the winds, including El Niño and La Niña events, are described as _____.

9. The deep ocean is considered a _____ because it removes carbon dioxide from the atmosphere.

10. _____ is the gradual physical or chemical process that breaks rock into smaller pieces.

11. Earth has experienced many _____ or large-scale disasters such as large volcanic eruptions or being struck by meteorites.

Name Date

Comprehension
Section 11.1

Use with textbook pages 467–475.

Factores that affect climate

1. What would be the temperature on Earth if the amount of greenhouse gases decreased?

2. What would be the effect on the climate in the northern hemisphere if the tilt of Earth increased from 23.5° to 24.5°?

3. Over time, the wobble in Earth's rotation will change. What effect will this have?

4. What is the relationship between the shape of the Earth's orbit and solar radiation?

5. What effect does an increase in yearly temperatures have on climate?

6. What is the main problem caused by melting glaciers?

7. What would happen to Earth's temperature if the levels of carbon dioxide released into the atmosphere continues to increase?

8. What are some of the effects of a volcanic eruption that could affect climate?

Use with textbook pages 476–479.

El Niño and La Niña

1. Answer the questions using the figures below.

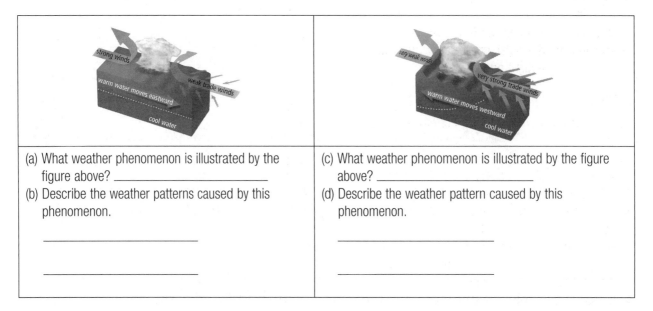

(a) What weather phenomenon is illustrated by the figure above? _____ (b) Describe the weather patterns caused by this phenomenon. _____ _____	(c) What weather phenomenon is illustrated by the figure above? _____ (d) Describe the weather pattern caused by this phenomenon. _____ _____

2. Using the weather maps of North America below, answer the following questions.

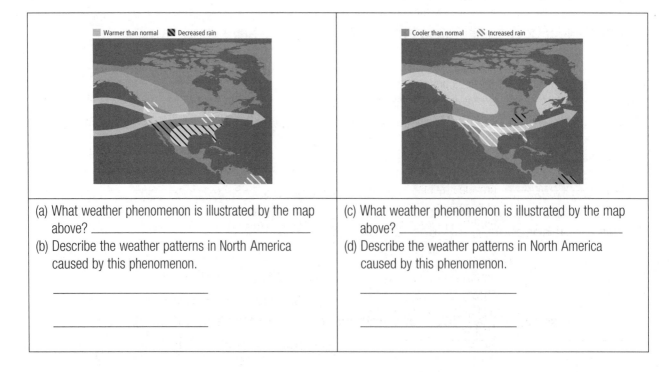

(a) What weather phenomenon is illustrated by the map above? _____ (b) Describe the weather patterns in North America caused by this phenomenon. _____ _____	(c) What weather phenomenon is illustrated by the map above? _____ (d) Describe the weather patterns in North America caused by this phenomenon. _____ _____

Use with textbook pages 464–479.

Natural causes of climate change

Match each Term on the left with the best Descriptor on the right. Each Descriptor may be used only once.	
Term	**Descriptor**
1. _____ carbon sink 2. _____ carbon source 3. _____ El Niño 4. _____ El Niño-Southern Oscillation 5. _____ greenhouse gases 6. _____ La Niña 7. _____ natural greenhouse effect 8. _____ paleoclimatologists	**A.** a body or process that releases carbon dioxide into the atmosphere **B.** a system of ocean and atmospheric changes in the tropical Pacific Region **C.** cooler-than-normal water coming to the surface in the eastern Pacific Ocean due to upwelling **D.** a body or process that removes carbon dioxide from the atmosphere and stores it **E.** an unusually warm ocean current that develops periodically off the coast of Ecuador and Peru **F.** people who study climates of the past **G.** the closed system, provided by the atmosphere, that keeps Earth's temperatures within a range **H.** gases in Earth's atmosphere that absorb and trap radiation as thermal energy

Circle the letter of the best answer.

9. Ice core data have been used to measure the amounts of which type of gas?

   A. oxygen   **C.** carbon monoxide

   B. nitrogen   **D.** carbon dioxide

10. An increase in greenhouse gases in the atmosphere will

    A. decrease temperatures on Earth

    B. increase temperatures on Earth

    C. make temperatures fluctuate

    D. have no effect on Earth's temperature

11. Which of the following are factors that affect the path of surface water currents?

    I. wind

    II. Earth's rotation

    III. shape of continents

    A. I only   **C.** I and III only

    B. I and II only   **D.** I, II, and III

12. An El Niño event results in

    A. cool temperatures in British Columbia

    B. cool temperatures in California

    C. warm temperatures in British Columbia

    D. warm temperatures in California

13. The remains of ancient marine organisms are composed of

    A. carbon dioxide   **C.** calcium carbonate

    B. sulphur dioxide   **D.** sodium carbonate

14. Catastrophic events, such as large volcanic eruptions, can affect climate by

    A. increasing the temperature of the troposphere

    B. decreasing the temperature of the troposphere

    C. decreasing carbon dioxide emissions

    D. increasing carbon dioxide emissions

Human Activity and Climate Change

Textbook pages 482–501

Before You Read

How might climate change affect the region where you live? Record your thoughts in the lines below.

_____

_____

What are climate change and global warming?

Climate change refers to changes in long-term weather patterns in certain regions. These changes affect the redistribution of thermal energy around Earth. **Global warming** describes an increase in Earth's average global temperature. It is one aspect of climate change. As greenhouse gases increase, the atmosphere is able to absorb and emit more thermal energy. This is known as the **enhanced greenhouse effect. Global warming potential (GWP)** describes the ability of a substance to warm the atmosphere by absorbing and emitting thermal energy. The table below shows the GWP of various greenhouse gases. The greatest carbon source resulting from human activity is fossil fuel combustion. Water vapour accounts for approximately 65% of greenhouse gases, carbon dioxide 25%, and other gases 10%. Chlorofluorocarbons are thought to be the main cause of the depletion of Earth's protective ozone layer. Humans have very little effect on the amount of water vapour in the atmosphere. Ozone, while an important greenhouse gas, is continually forming and breaking down, so it is difficult to determine its global warming potential.

Mark the Text

Check for Understanding

As you read this section, stop and reread any parts that you do not understand. Highlight any sentences that help you understand the concepts better.

Table 11.1 Green House Gases and Global Warming Potential				
Greenhouse Gas	Chemical Formula	Atmospheric Lifetime (years)	Source from Human Activity	Global Warming Potential (GWP)
carbon dioxide	CO_2	variable	• combustion of fossil fuels • deforestation	1
methane	CH_4	about 12	• combustion of fossil fuels • livestock agriculture • waste dumps • rice paddies	25
nitrous oxide	N_2O	114	• production of chemical fertilizers • burning waste • industrial processes	298
chlorofluorocarbons (CFCs)	various	45	• liquid coolants • refrigeration • air conditioning	4750–5310

Source: Intergovernmental Panel on Climate Change 2007

How do GCMs model climate?

General circulation models (GCMs) are computer models designed to study climate. They take into account multiple factors, such as changes in greenhouse gas concentrations, ocean currents, winds, surface temperatures, and albedo. The **albedo** at Earth's surface affects the amount of solar radiation that a region receives. GCMs are able to determine both past and present climate. Some models predict a temperature rise of 6°C in northern regions and a sea level rise of almost 88 cm within the next 100 years.

The effects of global warming may be most severe in northern countries, such as Canada. GCMs predict heavier spring rains and longer heat waves in some parts of the country. These changes will affect biomes across Canada as well as important industries, such as fisheries and forestry. Water resources and the health of Canadians may also be affected. Most regions of British Columbia will also become warmer. Some GCMs predict a 30 cm rise in the sea level along the northern coast of British Columbia over the next century. This could result in serious flooding in coastal communities. Areas of **permafrost**, ground that usually remains frozen year-round, are melting. The ice cover in the Arctic Ocean is rapidly shrinking.

How are governments addressing climate change?

The Intergovernmental Panel on Climate Change (IPCC) was established to address global concern about climate change and global warming. Its goal is to assess evidence of the human influence on climate change and suggest possible ways to respond. To encourage countries to reduce greenhouse gas emissions, the United Nations has set up an international environmental treaty called the United Nations Framework Convention on Climate Change (UNFCCC). As part of the treaty, countries determine what greenhouse gas emission limits should be. Because predictions about climate change cannot be certain, the United Nations suggests that governments use the **precautionary principle** to guide their responses to climate change. This principle states that a lack of complete scientific certainty should not postpone cost-effective measures to prevent serious environmental damage. ✓

To reduce the amount of greenhouse gases that Canada produces, the Canadian government has reduced allowable emissions from cars and trucks, required some industries to reduce emissions, increased the types of energy-efficient products available, and set guidelines for improving indoor air quality.

✓ *Reading Check*

What is the role of the Intergovernmental Panel on Climate Change (IPCC)?

_____

_____

_____

_____

Use with textbook pages 482–496.

Climate Change

1. Give three examples that illustrate that the Earth is undergoing a change in climate.

_____

_____

2. List the greenhouse gases that are produced by human activity.

_____

3. How is nitrous oxide formed?

_____

_____

4. What is the main cause of the depletion of Earth's protective ozone layer?

_____

5. What does the term albedo mean?

_____

_____

6. When computer general circulation models (GMCs) are designed to study climate, what factors are taken into account?

_____

_____

7. What evidence illustrates that northern Canada is being affected by global warming?

_____

_____

8. What plans have been implemented by the Canadian government to reduce our greenhouse gases?

_____

_____

_____

Use with textbook pages 484–486.

Greenhouse gases

The current increase in global temperature is caused by an increase in greenhouse gas emissions. Scientists have identified several produced by human activities.

1. Fill in the diagram below. Label the greenhouse gases and their approximate percent contributions to the greenhouse effect.

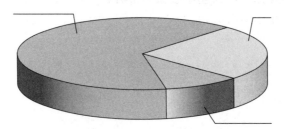

2. Complete the following table comparing greenhouse gases, their formulas, sources from human activity, and global warming potential (GWP).

Greenhouse gas	Chemical formula	Source from human activity	Global Warming Potential (GWP)
carbon dioxide			
methane			
nitrous oxide			
chlorofluorocarbons (CFCs)			

3. Explain why water vapour and ozone are not included on this table.

_____

_____

_____

Use with textbook pages 494–499.

Strategies for addressing climate change

Greenhouse gas concentrations in the atmosphere will increase during the next century unless greenhouse gas emissions decrease substantially from present levels. What can you do to reduce greenhouse gas emissions?

1. With a partner or small group, brainstorm ideas on how to reduce greenhouse gas emissions in your local community.

Brainstorm ideas:

2. What can you do personally to reduce your carbon footprint?

Use with textbook pages 482–496.

Human activity and climate change

Match each Term on the left with the best Descriptor on the right. Each Descriptor may be used only once.	
Term	**Descriptor**
1. _____ climate change 2. _____ enhanced greenhouse effect 3. _____ general circulation models 4. _____ global warming 5. _____ global warming potential 6. _____ permafrost 7. _____ precautionary principle	**A.** ground that remains frozen year-round **B.** the increase in global average temperature **C.** changes in long-term weather patterns in certain regions **D.** computer models designed to study the complex nature of climate **E.** the increased capacity of the atmosphere to trap thermal energy because of an increase in greenhouse gases **F.** the principle that a lack of complete scientific certainty should not be used as a reason to postpone cost-effective measures to prevent serious environmental damage **G.** the ability of a substance to warm the atmosphere by trapping thermal energy

Multiple Choice

Circle the letter of the best answer.

8. Computer models investigating global temperatures estimate that in one hundred100 years temperatures will increase by

   A. 1 °C

   B. 2.5 °C

   C. 6.0 °C

   D. 10.0 °C

9. Which gas contributes the most to the greenhouse effect?

   A. carbon dioxide

   B. chlorofluorocarbons

   C. methane

   D. water vapour

10. Which greenhouse gas is thought to be the main cause for the depletion of Earth's protective ozone layer?

    A. carbon dioxide

    B. chlorofluorocarbons

    C. methane

    D. nitrous oxide

11. Which of the following materials has the highest albedo?

    A. forests

    B. snow

    C. soil

    D. water

12. In Canada, greenhouse gas emissions come mostly from

    A. agriculture

    B. industry

    C. commercial heating

    D. transportation

Evidence for Continental Drift

Textbook pages 506–517

Before You Read

Scientists did not accept the continental drift theory for a long time. Why do you think this was the case? Write your ideas in the lines below.

_____

_____

 Mark the Text

In Your Own Words

After you read this section, explain the evidence for the continental drift theory in your own words.

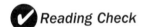 *Reading Check*

Millions of years ago, all the continents were joined as a "supercontinent." What was it called?

_____

What is continental drift?

In the early 20th century, German scientist Alfred Wegener proposed the **continental drift theory**, which argues that the continents "drifted" to their present locations over millions of years. On a world map, the curves of South America's eastern coastline and Africa's western coastline seemed to match, giving Wegener his first piece of evidence for continental drift. The fit suggested that, millions of years ago, all the continents were joined as a "supercontinent" named Pangaea (from the Greek words *pan*, meaning all, and *gaea*, meaning Earth). Wegener also noted that regions of some continents that are far apart have similar rocks, mountain ranges, fossils, and patterns of **paleoglaciation** (evidence of ancient glaciers and the rock markings they left behind). ✔

How do continents move?

After Wegener's death, scientists discovered that the surface of Earth is broken into **tectonic plates**, large, movable slabs of rock that slide over a layer of partly molten rock. According to **plate tectonic theory**, when tectonic plates move across Earth's surface, they carry the continents with them. Many volcanoes and earthquake zones on a map reveal the boundaries between the plates. Chains of volcanic islands, such as the Hawaiian Islands, reveal where tectonic plates have passed over geological **hot spots**—areas where molten rock has risen to Earth's surface. This idea was first suggested by Canadian scientist J. Tuzo Wilson.

Name

Date

Section

12.1

Summary

continued

The process of **sea floor spreading,** first proposed by Harry Hess, provides a mechanism for continental drift. This process involves magma, molten rock from beneath Earth's surface. Because it is molten, magma is less dense than the surrounding rock. Thus, magma rises and breaks through Earth's crust in certain weak areas. One such place is a **spreading ridge**, a gap in the sea floor that is gradually widening as tectonic plates move apart. Magma cools and hardens as it intrudes into this gap, pushing older rock aside as it creates new sea floor. The largest of all spreading ridges, and the first one discovered, is the **Mid-Atlantic Ridge**, a mountain range running north to south down the length of the Atlantic Ocean.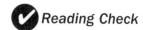

The evidence for sea floor spreading includes the following:

1. Age of ocean rocks: The youngest rocks are found closest to the ridge, indicating that new rock is being formed.

2. Sediment thickness: The layer of ocean sediment—the small particles of silt and organic debris deposited on the ocean floor—becomes thicker the farther it is from the ridge. This indicates that the sea floor is older and farther away from the ridge.

Reading Check

List one observation that provides evidence for sea floor spreading.

_____

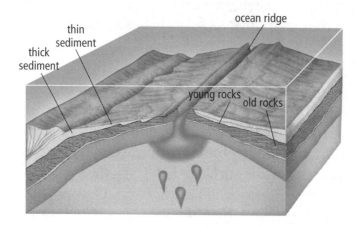

Name

Date

Section
12.1
Summary

continued

3. Magnetic striping: At a spreading ridge, iron-containing minerals in the magma align themselves with Earth's magnetic field as the magma cools. Because the orientation of Earth's magnetic field has switched many times over history, rocks on the sea floor exhibit both normal polarity and reverse polarity, depending on when they cooled. When scientists used a magnetometer, a device that detects variations in magnetic fields, they found a pattern of alternating polarity repeated on both sides of the Mid-Atlantic Ridge, as shown below.

This phenomenon, known as **magnetic striping**, indicates that new rock is being laid down on the sea floor. ✔

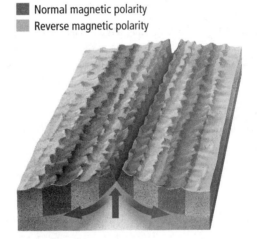

■ Normal magnetic polarity
■ Reverse magnetic polarity

Use with textbook pages 506–513

Evidence for continental drift

Vocabulary	
ancient glaciers	mountain ranges
fossils	Pangaea
geological structures	plate tectonic theory
hot spot	spreading ridge
magma	supercontinent
magnetic reversal	tectonic plates
Mid-Atlantic Ridge	

Use the terms in the vocabulary box to fill in the blanks. Each term may be used only once.

1. Alfred Wegener proposed that, millions of years ago, all the continents were joined as a _____

2. The name given to this giant land mass is _____

3. Wegener compared _____ ,
_____ and evidence of
_____ on different continents.

4. Since rocks found in Newfoundland are the same type and age as rocks found in Greenland, Ireland , Scotland, and Norway, it would appear that the world's major _____ were continuous when the continents were joined.

5. The surface of the Earth is broken into large, rigid, movable _____ that move over a layer of partly molten rock.

6. In the _____ , scientists found that as distance increases from the centre of the ridge, the rocks are older and the ocean sediment is thicker.

7. Using a magnetometer, scientists found a pattern of _____ in the iron-containing minerals on both sides of the Mid-Atlantic Ridge.

8. Harry Hess suggested that _____ rises because it is less dense than the material that surrounds it.

9. At a _____ the magma breaks through the Earth's surface, where it cools and hardens, forming a new sea floor.

10. J. Tuzo Wilson suggested that chains of volcanic islands were formed when a tectonic plate passed over a stationary _____.

11. The _____ is the unifying theory of geology.

Use with textbook pages 506–513.

Theories related to continental drift

Various pieces of evidence have been gathered by scientists to explain the underlying theories of geology. Alfred Wegener, Harry Hess, and J. Tuzo Wilson are some of the scientists who proposed explanations of phenomena they had observed.

Fill in the following table comparing the main points of evidence presented by each theory.

Continental drift	Paleomagnetism
Proposed by: _____ Main points: _____ _____ _____ _____	Main points: _____ _____ _____ _____ _____
Sea floor spreading	**Plate tectonic theory**
Proposed by: _____ Main points: _____ _____ _____ _____ _____ _____	Proposed by: _____ Main points: _____ _____ _____ _____ _____ _____

Use with textbook pages 509--515.

Visual observations supporting continental drift

Illustrations can demonstrate some of the major points related to the concepts presented in this chapter.

Refer to the diagrams on the left, when answering the questions below.

1.

Pangaea

What evidence did Wegener use for his explanations of the existence of Pangaea?

_____

_____

_____

_____

2. ■ Normal magnetic polarity
▢ Reverse magnetic polarity

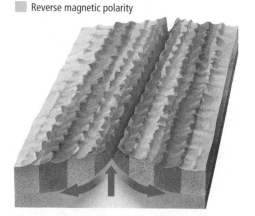

Orientation of Earth's Magnetic Field

(a) How were these magnetic patterns measured?

_____

_____

(b) What do these patterns show?

_____

_____

3.

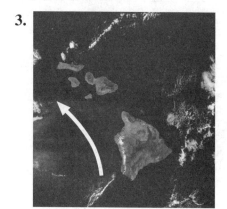

Hawaiian Islands

How were the Hawaiian Islands formed?

_____

_____

_____

Use with textbook pages 506–513.

Evidence for continental drift

Match each Term with the best Descriptor below. Each Descriptor may be used only once.

Term

1. _____ Continental drift theory
2. _____ Earthquakes
3. _____ hot spot
4. _____ magnetic reversal
5. _____ paleoglaciation
6. _____ plate tectonic theory
7. _____ spreading ridge
8. _____ tectonic plates
9. _____ volcanoes

Definition

A. the large slabs of rock that form Earth's surface and, move over a layer of partly molten rock

B. the theory that the lithosphere is broken up into large plates that move and then rejoin

C. an opening in Earth's surface that, when active, spews out gases, chunks of rock, and melted rock

D. an area where molten rock rises to Earth's surface

E. a pattern of alternating stripes of different directions of magnetic polarity in rock on the sea floor

F. a sudden, ground-shaking release of built-up energy at or under Earth's surface

G. the theory that the continents have not always been in their present locations but have moved over millions of years

H. the region where magma breaks through Earth's surface, continually forcing apart old rock and forming sea floor

I. the extent of ancient glaciers; also the rock markings they left behind

Multiple Choice

Circle the letter of the best answer.

10. When the term Pangaea is translated from Greek, it means

    A. zig-zag, continents

    B. stationary, plates

    C. all, Earth

    D. moving, plates

11. Which of the following would be considered part of Wegener's continental drift theory?

I.	Discovery of continents previously being part of a supercontinent.
II.	Matching fossils found on many continents.
III.	Discovery of reversal theories related to Earth's magnetic field.

    A. I only   **C.** I and III only

    B. I and II only   **D.** I, II, and III

12. A chain of volcanic islands, such as the Hawaiian Islands, were formed by which of the following processes?

    A. erosion   **C.** hot spots

    B. subduction   **D.** ocean ridges

13. J. Tuzo Wilson used which of the following to explain the theory of continental drift?

I.	sea floor spreading
II.	paleomagnetism
III.	formation of Pangaea

    A. I and, II only   **C.** II and, III only

    B. I and, III only   **D.** I, II, and III

Features of Plate Tectonics

Textbook pages 518–537

Before You Read

Earthquakes frequently occur in British Columbia. State what you already know about earthquakes in the lines below.

_____

_____

What lies below Earth's surface?

Earth is made of four layers with distinct characteristics. The **crust** is Earth's outermost layer. It is made from solid, brittle rock and is 5–70 km thick. The **mantle** is Earth's thickest layer. The upper mantle (660 km thick) is composed of partly molten rock that flows like thick toothpaste. A transition zone separates it from the lower mantle (about 2300 km thick), which is made of solid, dense material including magnesium and iron. Below the mantle lies the liquid **outer core** (about 2300 km thick and mostly nickel and iron). The **inner core** with a radius of 1200 km lies at Earth's centre. The incredible pressure at Earth's centre keeps the iron and nickel in the inner core solid. Scientists believe that the inner and outer cores rotate at different speeds, producing Earth's magnetic field. ✅

What are tectonic plates?

Tectonic plates are large, rigid, but mobile plates of rock. There are about 12 major tectonic plates and many smaller ones. Made up of the crust and the uppermost mantle, they form the **lithosphere**. Oceanic plates contain dense basalt rock. Continental plates contain large amounts of granite. Below the lithosphere lies the **asthenosphere**, a partly molten layer in the upper mantle.

Radioactive decay in some parts of the asthenosphere heats the mantle in these regions. Convection currents result as these hotter, and therefore less dense, regions of the mantle rise, cool, and sink again, only to be reheated. This **mantle convection** is one of the driving forces behind plate movement. ✅

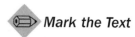

Identify Definitions

As you read this section, highlight the definitions of the words that appear in bold print.

Reading Check

List the four layers that make up Earth.

_____

_____

_____

_____

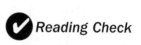

Reading Check

Name the process in the asthenosphere that pushes magma to Earth's surface, causing tectonic plates to move.

_____

How do tectonic plates interact?

A region where two tectonic plates are in contact is known as a **plate boundary**. The way in which tectonic plates interact depends on the type of plates and the direction the plates are moving relative to one another.

There are three main types of plate interactions:.

1. Divergence ⇌: **Divergent plate boundaries** mark the areas where tectonic plates are spreading apart. Such plates, for example, the East African Rift, are known as **diverging plates**. Rising currents of magma cool as they reach the surface and become "new" rock, resulting in spreading centres. A spreading centre in the ocean is called a spreading ridge or oceanic ridge. On land, it is called a **rift valley**. As new material at a ridge or rift pushes older material aside, the tectonic plates move away from the ridge. This process is called **ridge push. Rift eruptions** may occur when magma erupts at divergent plate boundaries. The Juan de Fuca ridge is an example.

2. Convergence ⣤ : A **convergent plate boundary** occurs where tectonic plates collide. Such plates are known as **converging plates**. If a dense oceanic plate collides with a continental plate, the heavy oceanic plate will dive under the lighter continental plate in an event known as **subduction**. A deep underwater valley, called a **trench**, forms where the plates make contact. As the edge of a tectonic plate subducts, it pulls the rest of the plate with it. This process is called **slab pull**. Along with convection currents and ridge push, slab pull helps keep tectonic plates in motion. As subduction occurs, magma can break through to the surface, forming volcanoes. A long chain of volcanoes, called a **volcanic belt**, may form as a result. The force of the collision between oceanic and continental plates also creates mountain ranges as the continental rock crumples and bends. British Columbia's Coast Mountains and Cascade Mountain Range were produced by such collisions.

Name

Date

Section

12.2

Summary

continued

Most volcanoes in volcanic belts are **composite volcanoes,** such as Mount Garibaldi, in British Columbia. Their cone shape results from repeated eruptions of ash and lava. **Shield volcanoes** are the world's largest, and their shape resembles a shield. They are not formed when plates collide, but when weaker areas of the lithosphere move over a "hot spot". The Anahim Belt is a chain of shield volcanoes in British Columbia. **Rift volcanoes**, like the Krafla volcano in Iceland, are formed when magma erupts through long cracks in the lithosphere.

Subduction also occurs where two oceanic plates converge. Cooling causes one plate to become denser. The denser plate slides deep into the mantle. Such convergence may produce a long chain of volcanic islands known as a **volcanic island arc.** The Aleutian islands and the islands of Japan are examples of a volcanic island arc. Subduction does not occur when two continental plates collide since the plates have similar densities. As continental plates collide, their edges fold, forming large mountain ranges, such as the Himalayas.

3. Transform ⇋ : Convection currents in the mantle often cause tectonic plates to slide past each other. Such regions are known as **transform plate boundaries**. Earthquakes and **faults** (breaks in rock layers due to movement on either side) may result. A fault that occurs at a transform plate boundary is known as a **transform fault**. The San Andreas Fault is an example of a transform plate boundary.

How are tectonic plates linked to earthquakes?

Friction between moving tectonic plates often works against convection currents, producing stress (the build-up of pressure). When the plates can no longer resist the stress, there is an earthquake—a massive release of energy that shakes the crust. The **focus** is the location inside Earth where an earthquake starts. The **epicentre** is the point on Earth's surface directly above the focus. An earthquake with a shallow focus (0–70 km) typically creates more damage than one with an intermediate focus (70–300 km), or a deep focus (greater than 300 km), as energy release occurs closer to the surface.

Energy released by an earthquake produces vibrations known as **seismic waves**. **Seismology** is the study of earthquakes and seismic waves. There are three types of waves: **primary waves** (**P-waves**) and **secondary waves** (**S-waves**), both of which travel underground, and **surface waves** (**L-waves**), which roll along Earth's surface.

Scientists use a seismometer to measure seismic waves. With each 1-step increase on the magnitude scale, the seismic waves are 10 times larger. Earthquakes can be felt if they are over magnitude 2.0. Over magnitude 6.0, they can damage building. P-waves are the fastest and stretch in the direction of the wave, like a spring. They can travel through solids, liquids, and gases. S-waves are slower, and move perpendicularly to the direction of the wave. They travel through solids but not liquids. L-waves are the slowest and cause a rolling motion like ripples on the ground.

Use with textbook pages 520--522.

Layers of the Earth

Earth is made up of layers with distinct characteristics.

1. Label the layers of the Earth on the following diagram.

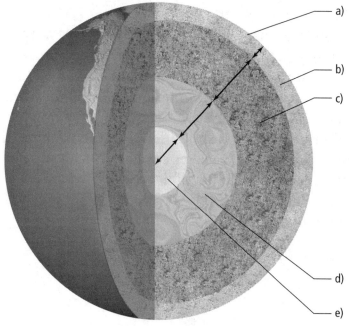

a)

b)

c)

d)

e)

Layers of the Earth

2. Each layer of the earth has a varying thickness, state (solid, liquid, gas) and composition. Fill in the following table beginning with the innermost layer in the order that you would find the layers from the inside to the outside of earth.

Layer	Thickness	State	General composition
(a)			
(b)			
(c)			
(d)			
(e)			

3. What is the difference between the lithosphere and the asthenosphere?

_____

_____

Name ______ Date ______

Comprehention
Section 12.2

Use with textbook pages 522-- 528

Features of plate tectonics

1. What do geologists believe heats the upper mantle portion of the asthenosphere?

______

2. What is one of the driving forces behind plate movement?

______

3. What is the difference between a rift valley and a spreading ridge?

______

4. What occurs when dense oceanic plates collide with a continental plate?

______

5. What events commonly occur at subduction zones?

______

6. When geologists record plate boundaries on a map, symbols are used to represent the three main types of plate interactions. Draw and label the three main symbols representing plate interactions.

(a) ______

(b) ______

(c) ______

7. Describe the type of plate interactions that have occurred at the following geographic locations.

Geographic location	Plate interaction
1. East African Rift	
2. Juan de Fuca plate	
3. Islands of Japan	
4, Himalayan mountains	
5. San Andreas Fault	

8. When continental plates collide, does subduction occur? Explain your answer.

______

______

Use with textbook pages 528–536.

Seismic waves, earthquakes, and volcanoes

Seismic waves can be either body waves that travel underground or surface waves that travel along the surface of the Earth.

1. Fill in the table below, summarizing the different types of seismic waves.

Seismic wave	Abbreviation	General diagram of wave	Description of action	Type of material it travels through	Speed it travels at
primary wave					
secondary wave					
surface wave					

Measurement of Earthquakes

2. What is a seismometer?

_____

3. How does the term magnitude relate to how earthquake activity is recorded?

_____

4. What scale is often used to measure the magnitude of an earthquake?

_____

5. What is the difference between the focus of an earthquake and the epicentre?

_____

6. Explain the classification scale used to describe the depth of origin of earthquakes.

_____

7. For the three geographic locations listed below, classify the type of volcano found there and describe what type of events led to its formation.

Geographic location	Type of volcano	Description of events
Mount Garibaldi volcano		
Anahim Volcanic Belt		
Kraflia volcano		

Use with textbook pages 518–534.

Features of plate tectonics

Matching

Match each Term on the left with the best Descriptor on the right. Each Descriptor may be used only once.

Term	Definition
1. _____ asthenosphere 2. _____ epicentre 3. _____ lithosphere 4. _____ mantle convection 5. _____ plate boundary 6. _____ ridge push 7. _____ rift valley 8. _____ slab pull 9. _____ subduction zone 10. _____ volcanic belt	**A.** the point on Earth's surface directly above the focus where an earthquake starts **B.** a recurring current in the mantle that occurs when hotter, less dense material rises, cools, and then sinks again **C.** the process in which new material at a ridge or rift pushes older material aside, moving the tectonic plates away from the ridge **D.** a steep-sided valley formed on land when magma rises to Earth's surface at a spreading centre on land **E.** a partly molten layer in Earth's upper mantle just below the lithosphere **F.** areas of subduction, which typically experience large earthquakes and volcanic eruptions **G.** a long chain of volcanoes **H.** the region where two tectonic plates are in contact

	I. the pulling of a tectonic plate as its edge subducts deep into the mantle **J.** the layer made up of the crust and uppermost mantle

Multiple Choice

Circle the letter of the best answer.

11. Which layer of the Earth has the highest temperature?

   A. inner core   **C.** lower mantle

   B. outer core   **D.** upper mantle

12. What causes the asthenosphere to be molten?

   A. gravity   **C.** hot spot

   B. volcanoes   **D.** radioactive decay

13. Where do transform plate boundaries usually occur?

   A. near mountains

   B. near continents

   C. near ocean ridges

   D. near subduction zones

14. Which layers of the earth can S-waves travel through?

   A. mantle only

   B. mantle and outer core

   C. mantle and inner core

   D. mantle, outer core, and inner core

15. Where are composite volcanoes usually found?

   A. near subduction zones

   B. in ocean basins

   C. on mountain ranges

   D. by ridge pushes

Glossary

This glossary provides the definitions of the key terms that are shown in boldface type in this workbook. The Glossary entries also show the sections where you can find the boldfaced words.

A

abiotic [ae-bih-AW-tik] relating to non-living parts of an environment such as sunlight, soil, moisture, and temperature (1.1)

acceleration the rate at which an object changes its velocity (9.1)

acceleration due to gravity (g) acceleration due to gravity in the absence of air resistance; the value of acceleration due to gravity near the surface of the Earth is approximately 9.8 m/s² downward (9.2)

acids compounds containing hydrogen that produce a solution with a pH of less than 7 when they dissolve in water and that produce a salt and water when they react with ionic compounds containing hydroxide ions (5.1)

adaptive radiation the development of a number of new species from a common ancestor; the new species are adapted to inhabit different niches (3.1)

air mass a large body of air with similar temperature and humidity throughout its length (10.2)

air resistance a friction-like force that opposes the motion of objects that move through the air (9.2)

albedo [al-BEE-doh] the amount of radiation reflected by a surface (10.2)

alcohol one kind of organic compound that contains C, H and O, such as methanol and ethanol (5.3)

alkali metal a very reactive metal, found in Group 1 on the periodic table (4.1, 5.2)

alpha decay the process in which an alpha particle is emitted from a nucleus (7.1)

alpha particle ($_2^4 \alpha$, $_2^4 He$) a positively charged atomic particle that is much more massive than either a beta particle or gamma radiation and is relatively slow moving; has same combination of particles as the nucleus of a helium atom (7.1)

angle of incidence the angle between a ray reaching a surface and a line perpendicular to that surface (10.2)

anions (AN-ih-uhnz) negative ions (4.1)

asthenosphere (uhs-THEN-uhs-feer) a partly molten layer in Earth's upper mantle just below the lithosphere (12.2)

atmospheres layers of gases that extend above a planet's surface (10.2)

atmospheric pressure the pressure exerted by the mass of air above any point on Earth's surface; also called air pressure (10.2)

atom the smallest particle of any element that retains the properties of the element (4.1)

atomic number the number of protons in the nucleus of an atom, which identifies the element to which the atom belongs (4.1)

average acceleration average rate at which an object changes its velocity; shown by the slope of a velocity-time graph (9.2)

average velocity the rate of change in position for a time interval (8.2)

B

balanced chemical equation a chemical equation that identifies each pure substance in the equation as well as shows the matching number of atoms of each element on both sides (4.3)

barometer an instrument used to measure atmospheric pressure (10.2)

bases chemical compounds containing hydroxide that produce a solution with a pH of more than 7 when they dissolve in water, and produce a salt and water when they react with ionic compounds containing positive hydrogen ions (5.1)

behavioural adaptations things that an organism does to survive in the unique conditions of its environment (1.1)

best-fit line a smooth curve or straight line that most closely fits the general shape outlined by the points on a graph; shows the trend of the data (8.1)

beta decay the process in which a neutron changes into a proton, which remains in the nucleus, and an electron, which is emitted from the nucleus along with energy (7.1)

beta particle ($_{-1}^{0}\beta, _{-1}^{0}e$) a high speed electron; emitted by a radioactive nucleus in beta decay (7.1)

binary covalent compound a compound that contains two non-metal elements joined together by one or more covalent bonds (4.2)

bioaccumulation the gradual build-up of synthetic and organic chemicals in living organisms (2.3)

biodegradation the breaking down of dead organic matter by living organisms such as bacteria (2.1)

biogeoclimatic zone a region with a certain type of plant life, soil, geography, and climate (11.1)

biomagnification the process in which chemicals not only accumulate but become more concentrated at each trophic level in a food pyramid (2.3)

biome [BIH-ohm] the largest division of the biosphere, which includes large regions with similar biotic components (e.g., similar plants and animals) and similar abiotic components (e.g., similar temperature and amounts of rainfall) (1.1)

bioremediation [bij-oh-re-mee-dee-AH-shuhn] the use of organisms—usually micro-organisms or plants—to break down chemical pollutants in water or soil to reverse or lessen environmental damage (2.3)

biotic relating to living organisms such as plants, animals, fungi, and bacteria (1.1)

Bohr diagram a diagram that shows the arrangement of an element's subatomic particles and the number of electrons in each shell surrounding the nucleus of an atom (4.1)

bonding pair a pair of electrons involved in a covalent bond (4.1)

bromothymol blue an acid-base indicator named after its colour change from yellow to blue over a pH range of 6.0 to 7.6 (5.1)

C

carbon cycle the nutrient cycle in which carbon is moved through the biosphere (2.2)

carbon sinks bodies or processes (e.g., plants, oceans, and soil) that remove carbon dioxide from the atmosphere and store it (11.1)

carbon sources bodies or processes (e.g., burning fossil fuels and trees) that release carbon dioxide into the atmosphere (11.1)

carbonate a combination of carbon and oxygen (CO_3^{-2}) that is dissolved in ocean water (2.2)

carnivores secondary consumers that eat primary consumers and often other secondary consumers. They are often at the tertiary level of a food chain. (2.1)

catalyst a substance that speeds up the rate of a chemical reaction without being changed or used up itself (6.2)

catastrophic events large-scale disasters (11.1)

cations [KAT-ih-uhnz] positive ions (4.1)

cellular respiration the process in which both plants and animals release carbon dioxide back into the atmosphere by converting carbohydrates and oxygen into carbon dioxide and water (2.2)

chain reaction an ongoing process in which one fission reaction initiates the next reaction (7.3)

change in velocity ($\Delta \vec{d}$) change that occurs when speed of an object changes, or its direction of motion changes, or both; calculated by subtracting the initial velocity from the final velocity (9.1)

chemical equation the representation of a chemical reaction in words or symbols (4.3)

chemical reaction one or more chemical changes that occur at the same time (4.3)

climate the average conditions of the atmosphere (e.g., precipitation, temperature, and humidity) in a large region over 30 years or more (1.1, 11.1)

climate change changes in long-term weather patterns in certain regions (11.2)

climatograph a graph of climate data for a region; the data are usually obtained over 30 years from local weather observation stations (1.1)

climax communities mature communities, such as a boreal forests, tropical rainforests, grasslands, or deserts, that continue to change over time (3.1)

coefficients numbers placed in front of a chemical symbol for an element that show the ratios between the various compounds in a chemical reaction (4.3)

combustion the rapid reaction of a compound or element with oxygen to form an oxide and to produce heat (6.1)

commensalism [kuh-MEN-suhl-ism] a symbiotic relationship in which one species benefits and the other species is neither helped nor harmed (1.2)

community all the populations of the different species that interact in a specific area or ecosystem (1.2)

competition a harmful interaction between two or more organisms that can occur when organisms compete for the same resource (e.g., food) in the same location at the same time (1.2)

composite volcanoes large, cone-shaped volcanic mountains; the cone shape is the result of repeated eruptions of ash and lava (12.2)

concentration the amount of substance dissolved in a given volume of solution (5.1)

conduction the transfer of thermal energy from one substance to another or within a solid by direct contact of particles (10.1)

conservation of mass the preservation of mass in a chemical reaction: the total mass of the product is always equal to the total mass of the reactants (4.3)

constant acceleration velocity changing at a constant rate (9.2)

consumers organisms that eat other organisms (2.1)

contamination the introduction of chemicals, toxins, wastes, or microorganisms into the environment in concentrations that are harmful to living things (3.2)

continental drift theory the theory that the continents have not always been in their present locations but have moved there over millions of years (12.1)

convection the transfer of thermal energy within a fluid and with the movement of fluid from one place to another (10.1)

convection current the movement of a fluid caused by density differences (10.1)

convergent plate boundary a region where tectonic plates are colliding (12.2)

converging plates tectonic plates that are colliding (12.2)

Coriolis effect [kor-ee-OH-lis} a change in direction of moving air, water, or objects due to Earth's rotation (10.2)

covalent bonding the formation of a chemical bond between atoms through the sharing of one or more pairs of electrons (4.1)

crust Earth's outermost layer formed by lighter materials, such as silicon and oxygen, floating to the top during Earth's early cooling period (12.2)

D

daughter isotope the stable product of radioactive decay (7.2)

DDT (dichloro-diphenyl-trichloroethane) [dih-KLOR-oh-dih-FEN-uhl trih-KLOR-oh-ETH-aen] an insecticide and well-known persistent organic pollutant, now banned in many countries (2.3)

decay curve a curved line on a graph that shows the rate at which radioisotopes decay (7.2)

deceleration acceleration that is opposite to the direction of motion; a decrease in the speed of an object (9.1)

decomposers organisms (e.g., bacteria and fungi) that break down wastes and dead organisms and change them into usable nutrients available to other organisms (2.1)

decomposition in biology, the breaking down of organic wastes and dead organisms (2.1); in chemistry, a chemical reaction in which a compound is broken down into two or more elements or simpler compounds (6.1)

deforestation the clearing or logging of forests without replanting (3.2)

denitrification [DEE-nih-tri-fi-KAY-shuhn] the process in which nitrogen is returned to the atmosphere (2.2)

denitrifying bacteria bacteria that convert nitrate (NO_3) back into nitrogen gas (N_2) (2.2)

detrivores consumers that feed at every trophic level, obtaining their energy and nutrients by eating dead organic matter (2.1)

displacement the straight-line distance and direction from one point to another (8.1)

distance *(d)* a scalar quantity that describes the length of a path between two points or locations (8.1)

divergent plate boundaries regions where tectonic plates are spreading apart (12.2)

double replacement describing a chemical reaction that usually involves two ionic solutions reacting to produce two other ionic compounds, either or both of which produce a precipitate (6.1)

E

earth metal a somewhat reactive metal, found in Group 2 on the periodic table (4.1, 5.2)

ecological hierarchy the order of biotic interactions and relationships in an ecosystem: organism, population, community, ecosystem (1.2)

ecological pyramid a food pyramid. There are three types of ecological pyramids: pyramids of biomass, pyramids of numbers, and pyramids of energy. (2.1)

ecological succession changes that take place over time in the types of organisms that live in an area (3.1)

ecosystem a part of a biome in which abiotic components interact with biotic components (1.2)

El Niño [el NEEN-yoh] an unusually warm ocean current that develops periodically off the coast of Ecuador and Peru, often producing unusually mild weather along the coast of British Columbia and in eastern Canada (11.1)

energy flow the flow of energy from an ecosystem to an organism and from one organism to another (2.1)

enhanced greenhouse effect the increased capacity of the atmosphere to trap thermal energy because of an increase in greenhouse gases (11.2)

epicentre the point on Earth's surface directly above the focus where an earthquake starts (12.2)

eutrophication [YOO-tri-fi-KAY-shun] the process by which excess nutrients in aquatic ecosystems result in increased plant production and decay (2.2)

F

faults large breaks in rock layers (12.2)

focus in geology, the location inside Earth where an earthquake starts (plural: foci) (12.2)

food chains models that show the flow of energy from plant to animal and from animal to animal (2.1)

food pyramid a model that shows the loss of energy from one trophic level to another; often called an ecological pyramid (2.1)

food web a model of the feeding relationships within an ecosystem; formed from interconnected food chains (2.1)

foreign species introduced species (3.3)

front the boundary between two air masses (10.2)

G

gamma decay a process in which an isotope falls from a high energy state to a lower energy state, giving off a high energy gamma ray; the result of a redistribution of energy within the nucleus (7.1)

gamma radiation ($^0_0\gamma$) rays of high-energy, short-wavelength radiation emitted from the nuclei of atoms (7.1)

General Circulation Models (GCMs) computer models designed to study the complex nature of climate (11.2)

geologic uplift the process of mountain building in which Earth's crust folds and deeply buried rock layers rise and are exposed (2.2)

global warming the increase in global average temperature (11.2)

global warming potential (GWP) the ability of a substance to warm the atmosphere by trapping thermal energy (11.2)

gravity attractive force between two or more masses; causes objects to be pulled toward the centre of Earth (9.2)

greenhouse gases gases in Earth's atmosphere that absorb and trap radiation as thermal energy (e.g., carbon dioxide) (11.1)

H

habitat the place in which an organism lives (1.2)

habitat fragmentation the division of habitats into smaller, isolated fragments (3.2)

habitat loss the destruction of habitats that usually results from human activities (3.2)

half-life in biology, the time it takes for a living tissue, organ, organism, or ecosystem to eliminate one half of a substance that has been introduced into it (2.3); in physics, the time required for half the nuclei in a sample of a radioactive isotope to decay, which is a constant for any radioactive isotope (2.3, 7.2)

heat the transfer of thermal energy from an area or object of high thermal energy to an area or object of low thermal energy (10.1)

herbivores primary consumers that eat plants (2.1)

hot spots areas where molten rock rises to Earth's surface (12.1)

humidity a measurement that describes the amount of water vapour in air (10.2)

hydrocarbon an organic compound that contains only the elements carbon and hydrogen (5.3)

hydrogen ions electrically charged hydrogen atoms (H^+); can be produced when acids are dissolved in solution (5.1)

hydroxide ions negative ions of OH^-; can be produced when bases are dissolved in solution (5.1)

I

ice cores cylinders of ice drilled from thick glaciers to determine the types and amounts of gases that existed in the atmosphere when the ice was formed (11.1)

indigo carmine an acid-base indicator named after its colour change from blue to yellow over a pH range of 11.2 to 13.0 (5.1)

infrared radiation heat radiation (10.1)

inner core Earth's solid centre (12.2)

inorganic compounds refers to compounds that generally do not contain carbon; the few exceptions include carbon dioxide, carbon monoxide, and ionic carbonates (5.3)

insolation the amount of solar radiation that reaches a certain area (10.2)

introduced species plants, animals, or micro-organisms that are transported intentionally or by accident into regions in which they did not exist previously (3.3)

invasive species introduced organisms that can take over the habitats of native species or invade their bodies (3.3)

ions electrically charged particles created when atoms gain or lose electrons (4.1)

ionic bonding the bond that forms as a result of the attraction between positively and negatively charged ions (4.1)

ionic compound a compound that is composed of a positive ion and a negative ion (4.2)

isotopes different atoms of a particular element that have the same number of protons but a different number of neutrons (7.1)

J

jet streams narrow bands of fast flowing air moving west to east in the upper troposphere at boundaries between cold and warm air (10.2)

K

keystone species species (e.g., salmon) that can greatly affect population numbers and the health of an ecosystem (2.3)

Kilopascals (kPa) the SI unit that measures the vertical force of atmospheric pressure per unit area (10.2)

kinetic energy the energy of a particle or object due to its motion (10.1)

kinetic molecular theory the theory that all matter is composed of particles (atoms and molecules) moving constantly in random directions (10.1)

L

La Niña [lah NEEN-yuh] cooler-than-normal water coming to the surface in the eastern Pacific Ocean due to upwelling; as a result, winter temperatures are often unusually warm in southeastern North America and unusually cold in the northwest (11.1)

land use the ways in which we use land, such as for urban development, agriculture, industry, mining, and forestry (3.2)

leaching removal by water of substances that have dissolved in moist soil (2.2)

Lewis diagram a diagram that illustrates chemical bonding by showing only an atom's valence electrons and its chemical symbol (4.1)

light one form of radiation that is visible to humans (7.1)

lithosphere the layer made up of Earth's crust and uppermost mantle and ranging in thickness from 65 to 100 km (12.2)

litmus paper thin paper strips coated with litmus and used as an acid-base indicator, turning one colour when added to a base, and a different colour when added to an acid. (5.1)

lone pair a pair of electrons in an atom's valence shell that is not used in bonding (4.1)

M

magnetic striping a pattern of alternating stripes of different directions of magnetic polarity in rock on the sea floor (12.1)

mantle Earth's thickest layer, lying just below the crust and making up 70 percent of Earth's volume (12.2)

mantle convection a recurring current in the mantle that occurs when hotter, less dense material rises, cools, and then sinks again. This current is believed to be one of the driving forces behind tectonic plate movement. (12.2)

mass number the total number of protons and neutrons found in the nucleus of an atom (7.1)

metal oxide a compound containing a metal chemically combined with oxygen (5.2)

methyl orange an acid-base indicator named after its colour change from red to yellow over a pH range of 3.2 to 4.4 (5.1)

methyl red an acid-base indicator named after its colour change from red to yellow over a pH range of 4.8 to 6.0 (5.1)

Mid-Atlantic Ridge the longest mountain range on Earth, running north to south down the middle of the Atlantic Ocean (12.1)

mimicry an adaptation in which a prey animal mimics another animal that is dangerous or tastes bad (1.2)

multivalent describing the ability of an element to form ions in more than one way, depending on the chemical reaction it undergoes (4.1)

mutualism a symbiotic relationship between two organisms in which both organisms benefit (1.2)

N

native species plants and animals that naturally inhabit an area (3.3)

natural selection the process in which, over time, the best-adapted members of a species will survive and reproduce. This process makes change in living things possible. (3.1)

neutralization (acid-base) the chemical reaction that occurs when an acid and a base react to form a salt and water (5.2)

niches the special roles organisms play in an ecosystem, including the way in which they contribute to and fit into their environment (1.2)

nitrification the process in which ammonium ($NHNH_4^+$) is converted into nitrate (NO_3^-) (2.2)

nitrifying bacteria soil bacteria involved in two stages of nitrification. In the first stage, certain species convert ammonium (NH_4^+) into nitrite (NO_2^-); in the second stage, different species convert nitrite (NO_2^-) into nitrate (NO_3^-). (2.2)

nitrogen cycle the nutrient cycle in which nitrogen is moved through the biosphere (2.2)

nitrogen fixation the process in which nitrogen gas (N_2) is converted into compounds that contain nitrate (NO_3^-) or ammonium (NH_4^+) (2.2)

non-metal oxide a chemical compound that contains a non-metal chemically combined with oxygen (5.2)

nuclear charge the electric charge on an atom's nucleus, which can be determined by counting the number of protons (4.1)

nuclear equation a set of symbols that indicates changes in the nuclei of atoms during a nuclear reaction (7.3)

nuclear fission the splitting of a massive nucleus into two less massive nuclei, subatomic particles, and energy (7.3)

nuclear fusion a process in which two low-mass nuclei join together to make a more massive nucleus (7.3)

nuclear reaction the process in which an atom's nucleus changes by gaining or releasing particles or energy (7.3)

nuclear symbol the standard atomic symbol for an isotope, including the chemical symbol, atomic number, and mass number (7.1)

nutrient cycles the way nutrients are cycled in the biosphere; the continuous flows (exchanges) of nutrients in and out of stores (2.2)

nutrients substances such as the chemicals nitrogen and phosphorus that are required by plants and animals for energy, growth, development, repair, or maintenance (1.2); important components of nutrient cycles in the biosphere (2.2)

O

organic refers to almost all carbon-containing compounds; exceptions include carbon dioxide, carbon monoxide, and ionic carbonates (5.3)

outer core the layer below Earth's mantle (12.2)

overexploitation the use or extraction of a resource until it is depleted (3.2)

oxide a chemical compound that includes at least one oxygen atom or ion together with one or more other elements (5.2)

P

paleoclimatologists [pael-ee-oh-klih-muh-TAWL-uh-jists] people who study climates of the geological past (11.1)

paleoglaciation [pael-ee-oh-glae-see-AE-shuhn] the extent of ancient glaciers; also the rock markings they left behind (12.1)

parasitism a symbiotic relationship in which one species benefits and another is harmed (1.2)

parent isotope the isotope that undergoes radioactive decay (7.2)

parts per million (ppm) a measurement of chemical accumulation; 1 ppm means one particle mixed with 999 999 other particles (2.3)

PCBs (polychlorinated biphenyls) pah-lee-KLOR-i-nae-ted bih-FEN-uhls] synthetic chemicals containing chlorine that are used in the manufacture of plastics and other industrial products, become stored in the tissue of animals, and also persist in the environment (2.3)

period each row of elements in the periodic table (4.1)

permafrost ground that remains frozen year-round (11.2)

pesticide a general term for a chemical that is used to eliminate pests, such as an insecticide that kills insects and a herbicide that kills weeds (2.3)

pH indicators chemicals that change colour depending on the pH of the solution they are placed in (5.1)

pH scale a number scale for measuring how acidic or basic a solution is (5.1)

phenolphthalein a chemical compound that is colourless in acidic or slightly basic solutions but turns pink in moderately basic to highly basic solutions (5.1)

phosphorus cycle the nutrient cycle in which phosphorus is moved through the biosphere (2.2)

photosynthesis a process in which carbon dioxide enters the leaves of plants and reacts with water in the presence of sunlight to produce carbohydrates and oxygen; photosynthesis also occurs in some micro-organisms (2.2)

physiological adaptations physical or chemical events that occurs within the body of an organism and enable survival (1.1)

pioneer species organisms such as lichens and other plants that are the first to survive and reproduce in an area; these organisms change the abiotic and biotic conditions of an area so that other organisms can survive there (3.1)

plate boundary the region where two tectonic plates are in contact (12.2)

plate tectonic theory the theory that the lithosphere is broken up into large plates that move and then rejoin; considered the unifying theory of geology (12.1)

polyatomic ion a molecular ion that carries a charge and is composed of more than one type of atom joined by covalent bonds (4.2)

POPs (persistent organic pollutants) harmful compounds containing carbon that linger for many years in water and soil

population all the members of a particular species within an ecosystem (1.2)

position ($\vec{d}$) a vector quantity that describes a specific point relative to a reference point (8.1)

position-time graph a graph of an object's position during corresponding time intervals; time data are plotted on the horizontal axis (x-axis), and position data are plotted on the vertical axis (y-axis) (8.1)

potential energy the stored energy of an object or particle due to its position or state (10.1)

precautionary principle the principle that a lack of complete scientific certainty should not be used as a reason to postpone cost-effective measures to prevent serious environmental damage (11.2)

precipitate an insoluble solid that forms from a solution (6.1)

predation predator-prey interactions in which one organism (the predator) eats all or part of another organism (the prey) (1.2)

prevailing winds winds that are typical for a certain region (10.2)

primary consumer an organism in the second trophic level (e.g., grasshoppers and zooplankton), which obtains its energy by eating primary producers (2.1)

primary producer an organism in the first trophic level, such as plants and algae (2.1)

primary succession the development of new life in areas where no organisms or soil previously existed, such as on bare rock; the first organisms may be lichen spores carried by wind (3.1)

primary waves (P-waves) seismic body (underground) waves that travel at about 6 km/s through Earth's crust, causing the ground to move in the direction of the waves' motion (12.2)

producers organisms that produce food in the form of carbohydrates during photosynthesis (2.1)

products pure substances formed in a chemical change that have different properties from those of the reactants (4.3)

R

radiation high-energy rays and particles emitted by radioactive sources (7.1)

radiation budget Earth's balance of incoming and outgoing energy (10.2)

radioactive decay the process in which the nuclei of radioactive parent isotopes emit alpha, beta, or gamma radiation to form decay products (7.1)

radioactivity the release of high energy particles and rays of energy from a substance as a result of changes in the nuclei of its atoms (7.1)

radiocarbon dating determining the age of an object by measuring the amount of carbon-14 remaining in it (7.2)

radioisotopes isotopes that are capable of radioactive decay (7.1)

rate of reaction a measure of how quickly products form, or given amounts of reactants react, in a chemical reaction (6.2)

reactants pure substances that react in a chemical change (4.3)

resource exploitation resource use (3.2)

resource use the ways in which we obtain and use naturally occurring materials such as soil, wood, water, gas, oil, or minerals (3.2)

ridge push the process in which new material at a ridge or rift pushes older material aside, moving the tectonic plates away from the ridge (12.2)

rift valley a steep-sided valley formed on land when magma rises to Earth's surface at a spreading centre (12.2)

S

salts a class of ionic compounds that can be formed during the reaction of an acid and a base (5.2)

scalar a quantity that has only a magnitude (does not include direction) (8.1)

sea floor spreading the process in which magma rises to Earth's surface at spreading ridges and, as it continues to rise, pushes older rock aside (12.1)

secondary consumer an organism in the third trophic level (e.g., frogs and crabs), which obtains its energy by eating primary consumers (2.1)

secondary succession the reintroduction of life after a disturbance to an area that already has soil and was once the home of living organisms (3.1)

secondary waves (S-waves) seismic body (underground) waves that travel at about 3.5 km/s, causing the ground to move perpendicular to the direction of the waves' motion; also known as shear waves (12.2)

seismic waves [SIHZ-mik] vibrations caused by energy released by an earthquake (12.2)

seismology the study of earthquakes and seismic waves (12.2)

shield volcanoes volcanoes that form over hot spots; the largest volcanoes on Earth (12.2)

single replacement describing a chemical reaction in which a reactive element (a metal or a non-metal) and a compound react to produce another element and another compound (6.1)

skeleton equation an equation that shows only the formulas of the reactants and products (4.3)

slab pull the pulling of a tectonic plate as its edge subducts deep into the mantle (12.2)

slope the direction of a line on a graph, either horizontal (zero), slanting up (positive), or slanting down (negative).

Slope is calculated by determining the ratio of rise/run. (8.1)

soil compaction the squeezing together of soil particles so that the air spaces between them are reduced (3.2)

soil degradation damage to soil—for example, as a result of deforestation or the removal of topsoil from bare land by water and wind erosion (3.2)

solar radiation the transfer of radiant energy from the Sun (10.1)

species a group of closely related organisms that can reproduce with one another (1.2)

speed (v) the distance an object travels during a given time interval divided by the time interval (8.2)

spreading ridge the region where magma breaks through Earth's surface, continually forcing apart old rock and forming new sea floor (12.1)

stable octet the arrangement of eight electrons in the outermost shell of an atom (4.1)

states of matter the properties of a substance describing it as a gas, liquid, or solid; may be shown in a chemical equation by the letters (g) for gas, (*l*) for liquid, (s) for solid, and (aq) for aqueous (dissolved in water) (4.3)

stores nutrients that are accumulated for short or long periods of time in Earth's atmosphere, oceans, and land masses (2.2)

structural adaptations physical features of an organism's body having specific functions that contribute to the survival of the organism (1.1)

subatomic particles the particles that make up an atom (4.1)

subduction the action of one tectonic plate pushing underneath another (12.2)

surface area the measure of how much area of an object is exposed; can affect reaction rate (6.2)

surface waves (L-waves) seismic waves that ripple along Earth's surface (12.2)

sustainability the ability of an ecosystem to sustain ecological processes and maintain biodiversity over time; using natural resources in a way that maintains ecosystem health now and for future generations (3.2)

symbiosis the interaction between members of two different species that live together in a close association (1.2)

symbolic equation a set of chemical symbols and formulas that identify the reactants and products in a chemical reaction (4.3)

synthesis a chemical reaction in which two or more reactants (A and B) combine to produce a single product (AB); also called a combination reaction (6.1)

T

tectonic plates the large slabs of rock that form Earth's surface, moving over a layer of partly molten rock (12.1)

temperature a measure of the average kinetic energy of all the particles in a sample of matter (10.1, Science Skill 7)

tertiary consumer [TUHR-shuh-ree] an organism in the fourth trophic level (e.g., hawks and sea otters), which obtains its energy by eating secondary consumers (2.1)

thermal energy the total energy of all the particles in a solid, liquid, or gas (10.1)

time interval (Δt) the difference between the final time and the initial time (when the event began) (8.1)

traditional ecological knowledge ecological information, passed down from generation to generation, that reflects human experience with nature gained over centuries (3.2)

transform fault a fault that occurs at a transform plate boundary (12.2)

transform plate boundaries areas where two tectonic plates slide past each other (12.2)

trench a deep underwater valley that is formed when an oceanic plate collides with a continental plate and is forced to slide beneath it (12.2)

trophic level a step in a food chain that shows feeding and niche relationships among organisms (2.1)

U

uniform motion travelling in equal displacements in equal time intervals; neither speeding up, slowing down, nor changing direction (8.1)

uptake the process of a substance, for example, nitrogen, entering plant roots and being incorporated into plant tissues (2.2)

V

valence electrons the electrons in the valence shell of an atom (4.1)

valance shell the outermost shell that contains electrons (4.1)

vector a quantity that has both a magnitude and a direction (8.1)

velocity ($\Delta \vec{v}$) the displacement of an object during a time interval divided by the time interval (8.2)

velocity-time graph a graph of an object's velocity during corresponding time intervals; time data are plotted on the horizontal axis (*x*-axis) and velocity data are plotted on the vertical axis (*y*-axis) (9.2)

volcanic belt a long chain of volcanoes (12.2)

volcanic island arc a long chain of volcanic islands (12.2)

W

water cycle the system of water circulation on, above, and below Earth's surface (11.1)

weather the condition of the atmosphere in a specific place and at a specific time (10.2)

weathering the process in which rock is broken down into smaller fragments (2.2)

wind the movement of air from an area of higher pressure to an area of lower pressure (10.2)

Periodic Table of the Elements

Legend / key box:

Atomic number	→ 22
Symbol	**Ti**
Name	Titanium
Average atomic mass**	47.9
	4+ 3+ → Ion charge(s)

metal

metalloid

non-metal

☐ O	natural
☐ Db	synthetic

Group 1

1	+
H	
Hydrogen	
1.0	

Period	1	2	3	4	5	6	7	8	9	10	11	12	13	14	15	16	17	18
1	1 + **H** Hydrogen 1.0																	2 0 **He** Helium 4.0
2	3 + **Li** Lithium 6.9	4 2+ **Be** Beryllium 9.0											5 3+ **B** Boron 10.8	6 **C** Carbon 12.0	7 3– **N** Nitrogen 14.0	8 2– **O** Oxygen 16.0	9 – **F** Fluorine 19.0	10 0 **Ne** Neon 20.2
3	11 + **Na** Sodium 23.0	12 2+ **Mg** Magnesium 24.3											13 3+ **Al** Aluminum 27.0	14 **Si** Silicon 28.1	15 3– **P** Phosphorus 31.0	16 2– **S** Sulphur 32.1	17 – **Cl** Chlorine 35.5	18 0 **Ar** Argon 39.9
4	19 + **K** Potassium 39.1	20 2+ **Ca** Calcium 40.1	21 3+ **Sc** Scandium 45.0	22 4+ 3+ **Ti** Titanium 47.9	23 5+ 4+ **V** Vanadium 50.9	24 3+ 2+ **Cr** Chromium 52.0	25 2+ 3+ 4+ **Mn** Manganese 54.9	26 3+ 2+ **Fe** Iron 55.8	27 2+ 3+ **Co** Cobalt 58.9	28 2+ 3+ **Ni** Nickel 58.7	29 2+ 1+ **Cu** Copper 63.5	30 2+ **Zn** Zinc 65.4	31 3+ **Ga** Gallium 69.7	32 4+ **Ge** Germanium 72.6	33 3– **As** Arsenic 74.9	34 2– **Se** Selenium 79.0	35 – **Br** Bromine 79.9	36 0 **Kr** Krypton 83.8
5	37 + **Rb** Rubidium 85.5	38 2+ **Sr** Strontium 87.6	39 3+ **Y** Yttrium 88.9	40 4+ **Zr** Zirconium 91.2	41 3+ 5+ **Nb** Niobium 92.9	42 2+ 3+ **Mo** Molybdenum 95.9	43 7+ **Tc** Technetium (98)	44 3+ 4+ **Ru** Ruthenium 101.1	45 3+ 4+ **Rh** Rhodium 102.9	46 2+ 4+ **Pd** Palladium 106.4	47 1+ **Ag** Silver 107.9	48 2+ **Cd** Cadmium 112.4	49 3+ **In** Indium 114.8	50 4+ 2+ **Sn** Tin 118.7	51 3+ 5+ **Sb** Antimony 121.8	52 2– **Te** Tellurium 127.6	53 – **I** Iodine 126.9	54 0 **Xe** Xenon 131.3
6	55 + **Cs** Cesium 132.9	56 2+ **Ba** Barium 137.3	57 3+ **La** Lanthanum 138.9	72 4+ **Hf** Hafnium 178.5	73 5+ **Ta** Tantalum 180.9	74 6+ **W** Tungsten 183.8	75 4+ 7+ **Re** Rhenium 186.2	76 4+ 3+ **Os** Osmium 190.2	77 3+ 4+ **Ir** Iridium 192.2	78 4+ 2+ **Pt** Platinum 195.1	79 3+ 1+ **Au** Gold 197.0	80 2+ 1+ **Hg** Mercury 200.6	81 1+ 3+ **Tl** Thallium 204.4	82 2+ 4+ **Pb** Lead 207.2	83 3+ 5+ **Bi** Bismuth 209.0	84 2+ 4+ **Po** Polonium (209)	85 – **At** Astatine (210)	86 0 **Rn** Radon (222)
7	87 + **Fr** Francium (223)	88 2+ **Ra** Radium (226)	89 3+ **Ac** Actinium (227)	104 **Rf** Rutherfordium (261)	105 **Db** Dubnium (262)	106 **Sg** Seaborgium (263)	107 **Bh** Bohrium (262)	108 **Hs** Hassium (265)	109 **Mt** Meitnerium (266)	110 **Ds** Darmstadtium (281)	111 **Rg** Roentgenium (272)	112 **Uub*** Ununbium (285)	113 **Uut*** Ununtrium (284)	114 **Uuq*** Ununquadium (289)	115 **Uup*** Ununpentium (288)	116 **Uuh*** Ununhexium (292)		

* Temporary names

Lanthanide series:

58 3+ 4+ **Ce** Cerium 140.1	59 3+ 4+ **Pr** Praseodymium 140.9	60 3+ **Nd** Neodymium 144.2	61 3+ **Pm** Promethium (145)	62 3+ 2+ **Sm** Samarium 150.4	63 3+ 2+ **Eu** Europium 152.0	64 3+ **Gd** Gadolinium 157.3	65 3+ 4+ **Tb** Terbium 158.9	66 3+ **Dy** Dysprosium 162.5	67 3+ **Ho** Holmium 164.9	68 3+ **Er** Erbium 167.3	69 3+ 2+ **Tm** Thulium 168.9	70 3+ 2+ **Yb** Ytterbium 173.0	71 3+ **Lu** Lutetium 175.0

Actinide series:

90 4+ **Th** Thorium 232.0	91 5+ 4+ **Pa** Protactinium 231.0	92 6+ 4+ 5+ **U** Uranium 238.0	93 5+ 3+ 4+ 6+ **Np** Neptunium (237)	94 5+ 3+ 4+ 6+ **Pu** Plutonium (244)	95 6+ 3+ 4+ 5+ **Am** Americium (243)	96 3+ **Cm** Curium (247)	97 3+ 4+ **Bk** Berkelium (247)	98 3+ 4+ **Cf** Californium (251)	99 3+ **Es** Einsteinium (252)	100 3+ **Fm** Fermium (257)	101 2+ 3+ **Md** Mendelevium (258)	102 2+ 3+ **No** Nobelium (259)	103 3+ **Lr** Lawrencium (262)

**Based on mass of carbon-12 at 12.00 u.

Any value in parentheses is the mass of the most stable or best known isotope for elements that do not occur naturally.